JIC	Just-in-Case
JIT	Just-in-Time
KISS	Keep It Simple, Stupid
LAN	Local Area Net
LED	Light-Emitting Diode
LSI	Large-Scale Integrated Circuit
LVDT	Linear Variable Differential Transformer
MAP	Manufacturing Automation Protocol
MBO	Management by Objectives
MIS	Management Information Systems
MODEM	Modulator–Demodulator
MRP	Materials Requirements Planning
MRP II	Manufacturing Resource Planning
MS-DOS	MicroSoft Disk Operating System
NBS	National Bureau of Standards
NC	Numerical Control
NEMA	National Electrical Manufacturers Association
OEM	Original Equipment Manufacturer
OSI	Open Systems Interface
PA	Purchasing Agent
PC	Programmable Controller or Personal Computer
PCB	Printed Circuit Board
PD	Programmable Device
PERT	Program Evaluation Review Technique
PMI	Plant-Manufacturing-Industrial Engineering
PROM	Programmable Read-Only Memory
QA	Quality Assurance
QC	Quality Control
R&D	Research and Development
RAM	Random-Access Memory
RF	Radio Frequency
RIA	Robotics Industries Association
ROM	Read-Only Memory
RTD	Resistance Temperature Detector
SAE	Society of Automotive Engineers
SCR	Silicon-Controlled Rectifier
SME	Society of Manufacturing Engineers
SNA	Systems Network Architecture
TC	Thermocouple
TIM	Token Interface Module
TTL	Transistor-to-Transistor Logic
VLSI	Very Large Scale Integrated Circuit
VP	Vice-President
WIP	Work In Process

Computer Integrated Manufacturing

THE PWS–KENT SERIES IN TECHNOLOGY

Computer Integrated Manufacturing, Vail
Digital Electronics: Logic and Systems, Third Edition, Kershaw
Electrical Systems in Buildings, Hughes
Electronic Communications Systems, Dungan
Electronic Instruments and Measurements, Crozier
Fundamentals of Digital Electronics, Knox
Fundamentals of Electronics, Second Edition, Malcolm
Fundamentals of Industrial Robots and Robotics, Miller
Industrial Electronics, Second Edition, Humphries and Sheets
Introduction to 6800/68000 Microprocessors, Driscoll
Introduction to Electronics Technology, Second Edition, McWane
Introduction to Electronics, Second Edition, Crozier
Linear Integrated Circuits for Technicians, Dungan
Microprocessor/Microcomputer Technology, Driscoll
Operational Amplifiers and Linear Integrated Circuits, Second Edition, Boyce
Principles of DC and AC Circuits, Second Edition, Angerbauer
Robotics: An Introduction, Second Edition, Malcolm
Technical Statics and Strength of Materials, Second Edition, Thrower

Computer Integrated Manufacturing

Peter S. Vail
New Hampshire Vocational Technical College
Manchester, New Hampshire

PWS–KENT Publishing Company **Boston**

20 Park Plaza
Boston, Massachusetts 02116

This book is dedicated to Fritter, my computer-aided writing machine.

PWS–KENT Publishing Company is a division of Wadsworth, Inc.

Library of Congress Cataloging-in-Publication Data

Vail, Peter S.
Computer integrated manufacturing / Peter S. Vail.
p. cm.
Bibliography: p.
Includes index.
ISBN 0-534-91465-9
1. Computer integrated manufacturing systems. I. Title.
TS155.6.V34 1988 87-20743
670.42′7—dc19 CIP

Printed in the United States of America
88 89 90 91 92 — 10 9 8 7 6 5 4 3 2 1

Sponsoring Editors: Robert Prior and George Horesta
Editorial Assistant: Mary Thomas
Production: Technical Texts, Inc./Betty O'Bryant
Interior and Cover Design: Sylvia Dovner
Typesetting: Publication Services, Inc.
Cover Printing: New England Book Components, Inc.
Printing and Binding: Halliday Lithograph

Acknowledgments: **Figures 1.2, 8.1, 8.3, 12.5, and 13.1:** Reprinted with permission from Ayers and Miller's *Robotics: Applications and Social Implications*, Copyright 1983, by Ballinger Publishing Company.
(Continued page 277)

▶ *Contents*

▶ *Preface*

Computer integrated manufacturing, or CIM, is the new buzzword in industry. Currently, the goal of CIM is the ability to tie together all functions of manufacturing into one large computer data base and to provide access to the data by various departments so that materials come in one door and products go out the other—all at the push of a button. However, CIM is somewhat of a misnomer since the ultimate goal is the ability to integrate all functions of business, not just manufacturing. So while we should be thinking about computer integrated business, we must crawl before we walk and at this point in time, can only explore the innovations and progress of CIM, keeping in mind that the principles are applicable to the larger concept of total business integration.

The design, installation, and maintenance of a CIM project in industry usually falls on the engineering department, and within engineering, the task frequently and rightfully falls on the industrial technology group. Unfortunately, almost no training exists in this new field, and so the practicing industrial technologist has to develop needed expertise on "the street," so to speak. Exposure to CIM should begin in IT departments at the college or university level. The intent of this book, then, is to provide access to this exciting and interesting subject for both students and practicing engineers. CIM is a fun field in which to be—one that has explosive growth potential and that will have wide sociological consequences.

The book treats new trends and developments in the field of CIM. It prepares the reader by first defining CIM, reviewing basic principles, and presenting a survey of available hardware, software, and interfaces.

The coverage includes "equipment" subjects, such as robots and robotics; CAE/CAD/CAM; CNC; types, sizes, and functions of mainframe, mini-, and microcomputers; visual and other feedback systems; data bases; networking; and software and hardware compatibility. Emphasis is also placed on "people" aspects, such as selection of a system; implementation; labor problems; middle management and human resources; and organizational structure. Interwoven throughout the book are advanced Japanese managerial and operational techniques that enhance productivity, including such concepts as Just-In-Time, elimination of Work-In-Process inventory, and inexpensive nontechnical techniques that can be carried out on the factory floor.

This text is a "how-to" book but on a more sophisticated level than most. On completing it, the reader should know what CIM is, how to plan a project for the incorporation of CIM, what kind of equipment is available and what it is supposed to do, how to put it all together, how to resolve the "thing" and "people" problems that CIM generates, and what the future holds for CIM and for the CIM engineer.

The audience for this book is the senior or top-ranked junior industrial technology major, graduate student, or practicing engineer in industry. Students who are thinking about what is taking place in industry have had most if not all of the prerequisites for a course in CIM and should be able to see CIM as a springboard toward career development. Courses that students have taken should include the following topics: operations management, quality control, statistics, computer programming and computer science, motion and time, controls and control systems, behavioral management, and engineering finance as well as the basics.

The book is intended as a one-semester comprehensive text. The thirteen chapters can be tucked into the following general format: introduction and review of basics, description of CIM units, putting all the technology together, sound management techniques, a look at human resource challenges, and, finally, thoughts on where we are headed. Chapter 5 can be skipped by students who have had a robotics course.

The Bibliography lists supplementary readings on data base management, general management, computer sciences, and related disciplines. Questions at the end of each chapter are intended to stimulate class discussions, while special student projects, such as a trade literature scavenger hunt, visits to nearby manufacturing firms, writing case histories of companies' successes and failures, are encouraged to add realism to the course. If time can be found, there are many opportunities for developing labs based, for example, on robotic components, microprocessors, networking with programmable devices, or game playing and computer simulations.

Many people made this book possible, the most important being my wife. She not only prodded me at key times but was my proofreader, grammar and style advisor, and sounding board. David Young, Special Sections Editor for the *Boston Globe* (retired) and Herb Haber, Professor, English Department, University of Lowell, also were of great help in the early stages of the writing. Technical help was rendered by Pat Krolack's group, particularly Brian Wheeler, at the University of Lowell Center for Productivity Enhancement, as well as by members of the IT department. Colleagues who reviewed the final draft of the manuscript and provided valuable comments include Robert Bittner, Trenton State College; William E. Pinebrook, Texas A&M University; Paul Dunham, University of Akron; John Stilt, Milwaukee Area Technical College; and Howard Terry Leeper, Western Kentucky University. The Bay States

Skills Corporation sponsored a CIM workshop whose mission was to train teachers, and it was a direct result of this workshop that I became interested enough in the educational need for CIM training to think about writing a book. My editor, George Horesta, and the people at Technical Texts, Inc., kept the ball rolling.

1 CIM: An Overview

In this chapter we introduce the concept of *computer integrated manufacturing*, more commonly known as CIM, and look at some of the reasons the study of CIM *today* is vital to competitive success *tomorrow*. CIM is a very powerful concept and has the potential to reap large benefits. However, like any powerful and complex tool, it can be a dangerous and costly tool if not applied properly.

1.1 Introduction

▶ ***CIM Is Growing Quickly.*** Computer integrated manufacturing (CIM) is an emerging technology. It is constantly changing. It is perhaps the fastest growing field today. CIM will create opportunities and problems for people who touch or are touched by CIM. Some of these changes can be anticipated, but some will come as a surprise. Therefore, it is essential the student have an open mind and be able to meet the challenge of change.

As in any growing technology, there is much overpromotion and exaggeration about the benefits of CIM. Buzz words abound like mosquitoes on a warm summer evening. CIM itself is such a buzz word. "It's the newest, it's the best," and it is "guaranteed" to cure all of industry's ills. Not so! CIM is a multifaceted tool that is useful in some applications and useless in others. Furthermore, like a carpenter's rafterframe square you have to know how to use it to get the most out of it.

▶ ***CIM Is a Management Philosophy.*** CIM is not a single tool. It is the whole tool box. All the technology, hardware, components, data structures, programming, and other individual tools used in a CIM environment are based on established technologies. There is nothing revolutionary to be found in CIM. What is different is that we as a society are now capable of putting it all together in one coherent system. CIM is the umbrella under which all the parts function. So, in reality, CIM is more of a philosophy of management operation than a technological discipline. The aim of this text is to show you what CIM is and how to use it. Whether you build a palace or a privy is up to you.

▶ ***CIM Is an Engineering Technology.*** Because CIM is an emerging field, there are no rules; we can provide only guidelines and few firm definitions. Since CIM is essentially a conglomeration of many technologies that have been financed by industrial concerns, governmental bodies, research and teaching institutions, and the like, many individuals and groups are involved, each with special interests or axes to grind. Everyone is attempting to promote a viewpoint, whether for prestige or profit. The problem for the student comes down to being able to separate the ballyhoo and fiction from fact.

For example, the Japanese classify robots differently than do Americans. Machines that look like and act like robots may not be classed as robots, and machines that are called robots by the maker may not fit a common definition of a robot. Classification of robots and what constitutes a robot are discussed more fully in Chapter 5.

Thus we must deal with an amorphous subject, which is something like trying to grab Jello. Knowing this probably won't decrease your anxiety, but it may help you to deal with those individuals who think their word is the only word. Since there are no clear definitions as to what constitutes the components of CIM, this text covers a broad range of topics that could be a part of CIM.

1.2 What Is CIM?

Components of CIM

CIM is an acronym for computer integrated manufacturing. Let's look at the parts of this phrase more closely. A computer is a machine that processes information. We put information into the machine, a transformation occurs, and we extract information from the machine. Presumably, the information we get out is more useful than what we put in. The computer can store information as it comes in, before it goes out, or anywhere in between. The computer can get information from many sources almost simultaneously and can output to many other machines, again almost simultaneously. If computers are designed with care, they can also exchange information with other computers.

The phrase "almost simultaneously" is a key phrase and we will talk about this later. For now, recognize that a computer is a sequential device. It gets a piece of information and does something with it. It does this very quickly and surely, which is why a computer is valuable; nevertheless, it processes one piece of information at a time.

▶ ***Managers and Computers.*** A manager receives information from a subordinate, transforms the information, and presents it to other managers and higher-ups, just like computers. Once upper management has digested the information, directives are issued and actions are effected. There are some things a manager can do better, faster, and cheaper than computers, and vice versa. The computer can also make comparative decisions and cause things to happen. Turning on a conveyor when the parts are ready is an example. The problem confronting CIM managers is deciding who or what is to do which to what or to whom.

▶ ***Concept of Integration.*** Integration means that all the computers and stored data are tied together. The stored data are called the data base or data-management system and may be at one central location or, more likely, spread around at several locations. Like control of voters in a big city, the control of data and who has access to what data are important questions. In some companies, the data are "owned" by the management information systems (MIS) group, and in others the data are "owned" by the department that generates and/or uses the data base. Regardless of who owns the data base(s), relevant data must be made available to whoever needs it in order for the system to be integrated. Some sort of communications network is required.

▶ ***What Manufacturing Does.*** Manufacturing is taking raw materials and transforming them into something else that has greater value. We make cars out of steel. With CIM, we use computers and a data base to help us make the cars out of the steel. Computers help us to design, order, and control the raw steel, operate the machines used to fabricate the cars, keep track of all aspects of production control, do scheduling, order and warehouse parts, make inspections and do many other important tasks. The computer also keeps track of costs of finished goods and work in process. Figure 1.1 is a model of a CIM configuration. Notice how the various operations such as assembly, manufacturing, and computer-aided design (CAD) are tied together by computer communication networks to the data base.

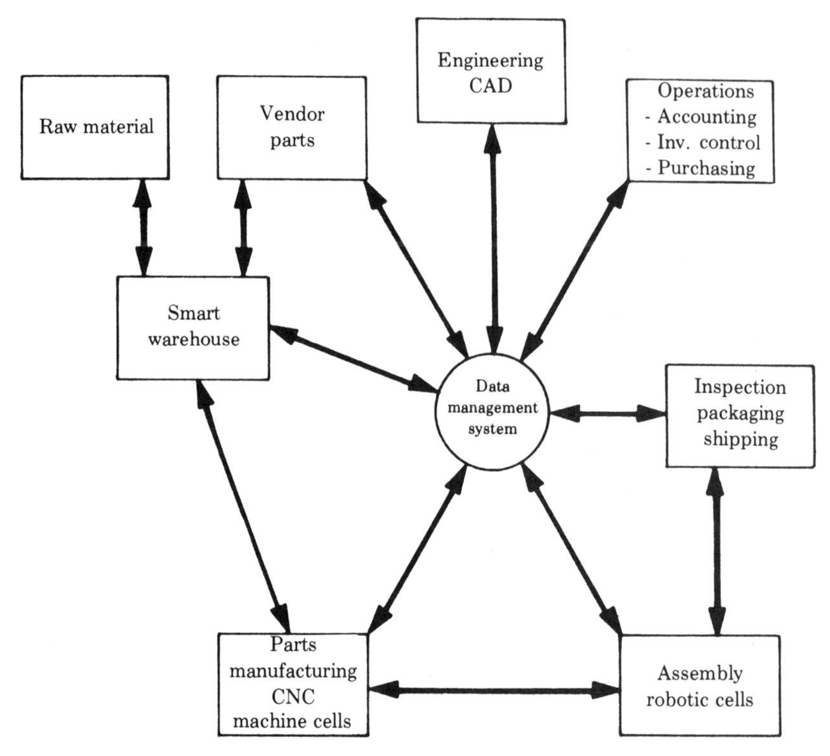

▶ ***Figure 1.1*** Computer Integrated Manufacturing

CIM versus CAD/CAM

CAM is an acronym for computer-aided manufacturing. Although CAM will be discussed in Chapter 7, it is important to note now that CAM uses the computer to drive the machine tool to make the part. Specific part design data input to the CAM unit computer often come from a CAD unit computer output. In this case, the CAD unit is integrated with the CAM unit. When the CAD function is integrated with the CAM function, it is designated CAD/CAM. Many people in industry use the term CAD/CAM interchangeably with the term CIM because the CAD/CAM function is vastly more developed than any of the other CIM functions and was the first of the various CIM functions. However, using the terms interchangeably is incorrect. CAD/CAM and other functions tied together with a computer are subsets of CIM. CIM and CAD/CAM are not synonymous terms. Although CAD/CAM is an important part of CIM, it is possible to have a manufacturing process without the CAD interface. An example would be a robot taught by a human expert to lay paint patterns for recall at a later time by a control computer.

▶ ***Expanding the Base of the CIM Concept.*** All the functions shown in Figure 1.1 and many more are associated with the manufacturing process. We can also use the computer to do market and sales forecasting, accounting and preparation of financial reports, personnel administration, and all other nonmanufacturing tasks, and we can integrate these functions into the data base. It is reasonable, then, to think of the concept in terms of computer integrated *management*, rather than the more limiting concept of computer integrated manufacturing. For many reasons, mostly economic, but some corporate, political, and secular as well, most companies start with the integration of manufacturing operations. However, the ultimate goal in any company should be to integrate the entire operation.

We speak of the integration of computers into manufacturing and of expanding this idea to encompass the entire business operation. Why not expand this concept to nonmanufacturing or service industries? Is there any reason why your school could not be fully computer integrated? Although the methods employed by CIM are, strictly speaking, manufacturing in origin and orientation, the implications in business and service industries are much broader.

1.3 The Goal of CIM

Synergistic Aspects of CIM

CIM is an idea whose time has come. To implement a CIM system means we have taken many technologies and blended them together so that the whole is better than the sum of the parts. This is called *synergism*. Synergism of existing and emerging technologies is the goal of CIM. Although we have identified the goal of CIM, this does not tell us much about why CIM is being considered in the first place.

▶ ***Nearly All Technologies for CIM Are Available.*** CIM has come into being because the technologies required to design and implement a CIM system are now available and an economic need has appeared to act as the driving force for this implementation. Quite simply, according to a product manager at a large firm, most companies implementing CIM are companies that are in pain due to economic loss. The Japanese have been able to build cars that are of higher quality than American cars and bring them to the United States for $2000 per car less. Part of the reason was because of more highly automated factories (CIM, robots, etc.), part because of better management of employees, and part because of cultural differences. Other countries besides the Japanese have eroded our competitive base. Many products in addition to cars have been affected, such as appliances, machine tools, cameras, and consumer electronics. Even in defense items we find foreign encroachment. Consider the U.S.

armed services who have adopted the Baretta handgun from Italy as their new standard, displacing the Colt 45. The driving force for CIM then is competition, mostly, but not exclusively, from abroad.

► ***Make Good Things Cheaply.*** To be competitive, we need to produce manufactured goods for less but still maintain adequate quality levels. A popular approach has been to reduce the amount of high-cost direct labor in finished goods. This is still a viable avenue in many industries, but in a large number of companies the ratio of direct labor (labor applied to the actual manufacturing cost of producing the product) to the total cost of the product is very low. In most companies, the largest single cost is material costs. Hence other places than direct labor need to be found for cost improvement. The most obvious is *work in process* (WIP), and we will explore this and the other areas. The hope and promise of CIM is that a proper system will reduce these costs.

1.4 What Is Manufacturing?

Manufacturing can be broken down into three basic types: continuous processing, custom or job shop and batch processing. These can be further subdivided, but these three major types will serve our purpose.

Continuous Processing

A continuous processing operation takes the raw materials and processes them into the finished product with very little, if any, work in process. An oil refinery or chemical plant is an example. These are characterized by very costly equipment and few operators. They are also highly instrumented in that 10% to 20% of the capital equipment is in the instrumentation. They are fully automated. The instruments are connected to the actuating mechanisms, such as valves, and so control the process. For some time, continuous processing has been computer controlled, and the computers are integrated. The terminology for CIM in these industries is *distributed systems*.

Custom or Job Shop Manufacturing

At the other end of the spectrum is the custom job shop or build-to-order company. Typically, this type of operation either uses very high technology and makes specialized products like gauges or is an artisan type of business, such as making custom executive furniture. Although these companies know how to use computers and could integrate the manufacturing, such a move is not cost justified. A high-technology, specialized company that does a great deal of engineering design would most likely use a computer to aid them in design and engineering, but not in the actual manufacturing. Computer-aided design is called CAD and computer-aided engineering is called CAE. Costs and pricing in the custom job shop reflect the expertise and talent of the principals, and there is not much need for CIM.

Batch Processing

The bulk of manufactured goods are fabricated on a batch basis. Each fabricated part is then assembled with other fabricated parts into subassemblies, which are further assembled, and so on. Note the distinction between fabrication of a part (e.g., cutting metal) and assembly of a group of parts. An assembly usually cannot begin until all the parts are on hand and ready. All fabricated parts remain as WIP until the last part is ready and the assembly is complete. At this point, the product passes to the finished goods status. It is desirable to reduce assembly time from 10 hours to 2 hours, for example, but it is absolutely necessary to reduce part fabrication from 14 weeks to 1 week. CIM finds widest potential applicability in a manufacturing environment that is producing products on a batch basis. Figures 1.2 and 1.3 show the continuum between continuous or mass, batch, and job shop or piece operations. Notice in Figure 1.3 that, as the product variety increases, it becomes more attractive to consider programmable automation and robotics.

Role of Automation

As we move to larger and larger batch sizes, it pays for us to use more and more hard automation. Hard automation refers to a piece of equipment dedicated to a particular part and process. A crankshaft fabrication line in an engine plant uses a great deal of hard automation, as does a beer bottling plant. Although these machines are fast and efficient, reprogrammability is not one of their hallmarks. Even so, tasks such as infeed and outfeed are usually under computer supervision. The line between hard and soft (computer controlled) automation is, at best, fuzzy. There are installations using computer-controlled robots for which the programs, machine settings, and tooling have not been changed in three years.

1.5 Characteristics of a System

CIM has incorrectly been referred to as a system or a grouping of subsystems. A system is defined by one dictionary as "a whole scheme of created things regarded as forming one complete whole." There are systems that have CIM components and there are companies that have gone very far in integrating their manufacturing plant. However, there

	Type of Process		
	Piece	***Batch***	***Mass***
Primary motivation	Ability	Flexibility	Volume and cost minimization
Part variety	Many different shapes and sizes	Mostly similar in shape, type of materials	Essentially identical parts with few processes and materials
Flexibility of machine tools	High—can combine several processes in one machine and vary each process	Medium—limited to part family, can change tools, speeds, feeds, and dimensions	Low—usually only minor changes in speeds, feeds, or part dimensions
Flexibility to change to a completely different part	Yes	Possible	Impossible
Flexibility to change to similar, but different part	Yes	Yes, if previously planned	Very limited
Flexibility to change materials	Yes	Limited	Extremely limited
Flexibility of plant layout	Smaller machine tools can be added	Some flexibility, but difficult to install new machines in system	Carefully planned, but not flexible; not usually possible to install new machines in system

▶ ***Figure 1.2*** Flexibility of Piece, Batch, and Mass-Production Systems (Ayers and Miller, *Robotics*)

is no such thing as a CIM system per se. Talk about CIM as a holistic system has the characteristics of a myth. It's like looking for the Holy Grail. We can talk about what the ideal is, but achieving it is something else. As stated earlier, CIM is an umbrella or philosophy under which technologies can function.

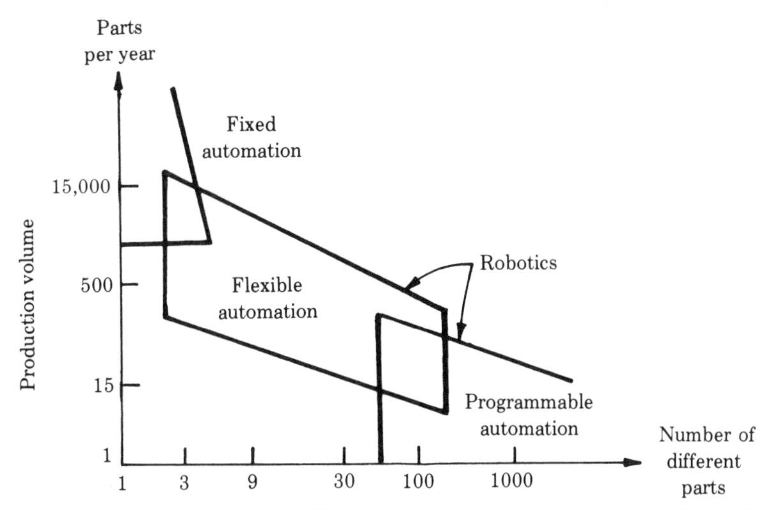

The chart shows the relationship of fixed automation, programmable automation, and flexible automation as a function of production volume and product variety.

▶ ***Figure 1.3*** Automation and Product Characteristics (Groover, *Industrial Robotics*)

▶ ***CIM Cannot Be a System.*** The subtle distinction between CIM as a system and CIM as an environment may seem trivial, but failure to grasp the difference can lead to very costly mistakes. What typically happens is that someone in upper management, who knows all the latest buzz words, decides to build a CIM plant or "Factory of the Future." The plant is built but does not work. A perfect example of this kind of unrealized optimism is the John Deere facility on which over $1 billion was spent. It is important to realize that the process of implementing CIM is evolutionary, and to build a CIM plant from scratch frequently leads to disappointment and disillusionment. To say "our CIM venture was a failure" is like blaming one's failure on the air.

1.6 Implementation by Decree: The Old Way

Importance of Management Methods

One reason the Japanese have been so successful is because there is a fundamental difference in the relationship between management and the worker compared with what we find in many U.S. factories. In this country we take the position that management specifies and designs. The worker is expected to carry out the directive. The problem with this approach is that when systems become complex the method simply does not work. There may be technical reasons on the shop floor. There may be political factors between departments that inhibit cooperation. Or, there may be motivational problems. Part of the situation traces back to the "us versus them" attitude found in many industries. Management usually is quick to blame the union and the worker. However, many people are coming to the conclusion that, if you treat a person or group of people as though they were stupid, incompetent, noncaring, and the like, this is the way they will behave. The relationship between the Japanese worker and management is much more open and trusting. For this reason, goals are established together, and the net result is things get done better and quicker. Not only does the Japanese worker have an opportunity to express an opinion, but he or she is frequently involved in the decision making as well.

One fear of relinquishing decision making to lower levels is loss of control in general and loss of control by management in particular. Incorporation of CIM can overcome this objection in an objective way by providing factual information against which performance can be measured. We will return to these ideas in Chapter 2.

1.7 Quality Assurance versus Quality Control

► ***Quality Assurance Is Better Than Quality Control.*** By involving the worker in the decision making, the worker is likely to have enough self-esteem to be willing to accept the responsibility for making good parts. You can tell a machine operator that he or she is responsible for making good parts and the worker may say "OK," but if the worker does not, in fact, accept the responsibility, you will not get good parts. Many factory managers do not know this. The result of this fundamental misunderstanding on the part of management is that to have good parts it is necessary to have massive quality-control programs. For parts to be made available for shipment, we need to make a lot of parts, inspect them, cull out the bad ones, and inventory them. However, if we make each machine operator his or her own chief inspector and the operator accepts this role, then good parts will be fabricated. Naturally, the operator must be well trained in quality matters to assure that parts are going to be acceptable, and the machinery used must be technically capable of producing good parts. One way to do this is to have the machine monitor the output and alert the operator when things are not right. Computers do this very well. Thus the idea is to build quality assurance into the system so that we can reduce or eliminate the need for quality control and third-party inspection.

1.8 Just-in-Time versus Just-in-Case

If, through the use of computer-integrated machinery and enlightened management, we can make good parts everytime, then we have taken a giant step toward eliminating work in process. The idea is to have the part available at the next stage of processing the instant that stage is ready for the part. This is called just-in-time (JIT) or, in Japanese, Kanban.

► ***Just-in-Case Is Traditional.*** The traditional method is called just-in-case (JIC). We make and keep parts just in case some don't work and we are unable to rely on the previous fabricator in the chain to deliver on time. Let's say we make special shafts. Ideally, we buy bar stock (raw material) and then begin to process it. We select the right diameter, cut it, turn it, grind it, slot it, drill it, tap it, and so on, and then ship it (finished goods). However, under JIC what really happens is at each stage of the process, the WIP goes to a holding area where it waits its turn for the next operation. If it is processed incorrectly, it must be reworked, scrapped, or (more frequently) tagged as below quality but passed anyway. In this fashion, it may take months to get the finished assembly out the door. The actual fabrication time at each stage may be relatively short, a matter of hours, but the delay times are usually days or weeks. Thus, the consumer of the shafts must wait an inordinate amount of time for the shafts or must stockpile excessive quantities to prevent stockout.

Important Aspects of JIT

Two requirements must be met if just-in-time is to work. First, the part that is delivered must be within acceptable tolerances. It has to pass the quality-assurance criteria discussed previously. The second requirement mandates that the part be delivered when promised. In a just-in-time environment, any delay in delivery is liable to shut down the entire operation. This simply must not happen. Not only is perfect scheduling implied, but so is perfect machine operation. In-depth analysis of machine function, tool requirements, procedures, design of finished parts, transport systems, and all other aspects of the system will lead to contingency planning. If a tool breaks, how is the replacement tool to get there so as to not interrupt production? Extensive gauging and physical control are required to compensate for tool wear.

1.9 Feedback Principle

Control Procedures

With just-in-time, we have to build in instant feedback. If the part you deliver to the next station is not right, it cannot be used. Since there is no WIP, the entire fabrication/assembly comes to a standstill. Everyone from the janitor to the chief executive officer's wife knows when production stops. How do we prevent a plant shutdown from happening? We design the process so that bad things cannot happen. For example, raw materials coming into the plant must meet specifications and arrive on time. Vendors must be trained or eliminated. The computer and data base in our purchasing department will need to be integrated with the computer and data base of our vendor. We can use on-line automatic test equipment (ATE). If a tool is showing wear, we can automatically change the tool with a programmable controller, a very small, dedicated computer. We need to schedule parts delivery so that they are at the right place at the right time, oriented properly, and in the correct sequence. Computers are consistent and do these tasks well if programmed properly. The programs have to be right.

1.10 Pareto's Law

Basically, what we are dealing with is a question of using the big guns on the big problems and letting the small things go. In a typical company, 80% of the dollar value of inventory will be represented by 20% of the goods. The relationship more or less fits a hyperbola. The underlying principle was first identified by Vilfredo Pareto, a social scientist, in the late 19th or early 20th century. The really expensive items, of which there are few, usually get the personal attention of an individual manager. It is then possible to use a computer to keep track of everything else, and many companies do just that. The problem with this approach is that keeping track of every item on the computer takes valuable memory and data-entry time and is therefore costly. A better way would be to select a dollar cutoff figure and keep enough of the cheap materials on hand so that you don't run out.

► ***Handle Inexpensive Items with Automatic Procedures.*** As an example, consider fasteners. First, fasteners, although cheap, lead to expensive assembly and should be avoided in the design. Even so, fasteners can be inventoried using the two-bin method. As soon as the primary bin is empty, a reorder is entered. Parts are then taken from the secondary bin. When the order comes in, the primary bin is filled. It now becomes the secondary bin and the secondary bin becomes the primary. The bins have to be large enough to prevent stock-out. This is a simple and cheap solution that obviates the computer. Similar thinking should be applied to CIM, but in many cases it is not. "We are going to computerize this plant" is what frequently comes down from on high. Millions of dollars later, we find machines in certain locations where direct labor would be cheaper. We find complex solutions to simple problems. We find things being checked that don't need it and things that should be checked that aren't. In short, management has lost sight of Pareto's law. The answer is to use CIM where CIM has value and recognize that CIM is not a cure-all.

1.11 Structure of the Book

This book is divided into four parts. The first part, which includes Chapters 1 and 2, introduces you to some basic concepts of CIM. Also included are why CIM is a viable technology today and a brief review of some of the fundamentals. The second part, which includes Chapters 3 to 6, describes the basic machine units of CIM. A point that will be emphasized throughout the text is that not all units will be used in all CIM systems. You can think of each individual unit as a component on a menu to be selected as desired. Starting with Chapter 7, we bring the discrete units into a system, putting it all together, so to speak. The third part deals with most of the problems that have already come up with the implementation of CIM. Like weeds, there is a whole field of problems waiting to pop up given the right conditions. When you move on to industry, your job will be to find and remove those weeds. In this text we will try to anticipate some of the problems and try to prepare you for the new ones that you will discover on your own.

1.12 Summary

CIM means computer-integrated manufacturing. The goal is to tie all the various computers, programmable controllers, and other programmable devices found in the factory into one integrated network or system wherever it is operationally advantageous and profitable to do so. Without the integration of the various machines and their associated computers, we have islands of automation. Since CIM is a grouping of technologies and since the goal is constrained by economic realities, we can think of CIM as a management philosophy that allows us to optimize our productive resources.

The reason why CIM is becoming so important is because without it we will continue to lose our competitive advantage both at home and abroad. Sensible implementation of CIM will allow us to produce high-quality goods at less cost.

To implement a CIM philosophy, we need to consider it as an environment under which programmable automation can flourish. There is little point in implementing CIM unless we precede it with sound management methods, such as a just-in-time system, and we maintain quality standards. Good control and procedures are essential for success.

1.13 Exercises

1. What is the definition of CIM?
2. What rules govern CIM?
3. Name the key limitation of a computer.
4. What do computers do that managers don't?
5. Who owns the data in a company?
6. Is CIM ready to be implemented? Discuss.
7. What technologies do we need to implement CIM?
8. What are the three kinds of manufacturing?
9. Where does CIM apply in the three kinds of manufacturing processing?
10. What is the difference between manufacturing and assembly?
11. What is a system?
12. What does reprogrammability mean?
13. What is the key difference between Japanese and U.S. management methods?
14. What is the difference between quality assurance and quality control?
15. What does just-in-time mean?
16. How does just-in-time compare with the traditional U.S. method and what is this method called?
17. What does the principle of feedback mean?
18. How do we prevent a plant shutdown in a just-in-time environment?
19. Describe the two-bin method.

2

Productivity and Work

2.1 Introduction

This chapter deals with how we traditionally look at and measure work, what is wrong with the traditional methods, and ways to improve the situation. We identify problems that occur not only on the shop floor but also in the executive suite. We even touch upon problems in our cost accounting methods.

▶ ***Paradox of Traditional Productivity Definitions.*** If we produce $1,000,000 worth of goods and employ 1000 people in direct manufacturing, we can say the productivity per employee is $1000. This is what is usually thought of as productivity. Consider the following: If we automate our factory somewhat and reduce the work force to 833, our productivity has increased to $1200. A reduction to 500 yields $2000. Let's automate completely and get rid of all the workers. Now the productivity is infinity!

Statistics from the Bureau of Labor Statistics, Department of Labor, and other governmental bodies bombard us with comparative productivity figures between the United States and other countries. Productivity in the United States is defined as GNP (Gross National Product) output divided by direct labor hours worked and is listed as an index, with 1977 being the base year. "The productivity of the American worker has dropped to so and so," is what we read, "whereas the Japanese worker's productivity has increased to such and such." Obviously, we need to rethink and redefine what we mean by productivity and work.

2.2 Definitions

One definition of productivity is: The ratio of outputs produced per unit of resources consumed compared to a similar ratio from some base period. This definition is broad enough to include just about everything, so we need to look at the terminology a little more closely.

Output

Outputs produced could mean dollar value of goods or number of goods. This definition would be applicable to a manufacturing operation, but not to a service. We have to be very careful to define output in terms that have real value. A lawyer might try 10 cases in one month and lose them all. A popular measure in education is the number of students graduated. But, did they learn anything?

Time Period

The time period is easier to come to grips with. Here we are dealing with a specific interval, say a month or year, compared with the same interval at some other period, called the *base period*. This then becomes an index when we divide the measured period by the base period. Care must be taken to use the same base period when comparing one ratio against another.

Input

As indicated, the most common measure of productivity is to use the number of workers (labor) or the number of labor hours as the input. However, we saw what happened when we applied this measure in the automated factory. Because of this situation, we have to be able to look at other measures of productivity. Some other productivities are based on direct labor cost, capital, total cost, foreign exchange, raw material, and energy, to name but a few. Figure 2.1 shows some of the measures of productivity that could be used. This quickly takes us into an Alice-in-Wonderland environment in which we can say, "Productivity is whatever I say it is!" Output per worker could be going up (half the work force got laid off), while output per value of capital used is going down (because we automated the plant). Human nature being what it is, a manager will use whatever definition suits his or her purpose at the time. Clearly, we need to come up with meaningful and more comprehensive measurements of the productivity of a CIM facility.

Kinds of productivity measures. Productivity measures may vary with respect to what aspect or aspects of both the outputs and inputs are used as a basis of aggregation. We may have:

Labor productivity index—In this formulation the resource inputs are aggregated in terms of labor hours. Hence, the index is relatively free of changes caused by wage rates and labor mix.

Direct labor cost productivity—In this formulation the resource inputs are aggregated in terms of direct labor costs. This index will reflect the effect of both wage rate changes and changes in the labor mix. However, *constant dollars* may be used to eliminate the distortion caused by inflation and deflation.

Capital productivity—Two formulations are possible: (1) the resource inputs may be the charges during the period to depreciation; (2) the inputs may be the book value of the capital equipment used.

Direct cost productivity—In this formulation, all items of direct cost associated with resources used are aggregated on a monetary value basis. Constant dollars may be used to avoid distortions from inflation or deflation.

Total cost productivity—In this formulation, all resource costs, including depreciation, are aggregated on a monetary basis. Constant dollars may be used.

Foreign exchange productivity—In this formulation, the only resource cost considered is the amount of foreign exchange required.

Energy productivity—In this formulation, the only resource cost considered is the amount of energy consumed, in Btu or kWh, as may be convenient.

Raw materials productivity—In this formulation, the numerators are usually the weight of product, the denominators the weight or value of raw materials consumed.

Hence, it may be seen that many different productivity indexes may be constructed. The list given is not exhaustive.

▶ ***Figure 2.1*** Productivity Measurements (Mundel, *Improving Productivity & Effectiveness*)

Define Costs Accurately

The productivity of a single multiproduct manufacturing facility is more difficult to measure than is the productivity of an entire industry. The problem with the single plant is in determining what costs should be included in the cost of each product and how to allocate overhead costs such as research and development. In the case of an entire industry, the allocation question disappears. In the automotive industry, we find comparisons can be made on a plant-wide basis for a specific type of car.

If the investment in a plant to make the same number of mid-sized cars is less in Japan and if that plant employs fewer *total* number of people than the American counterpart, then the productivity of the Japanese plant is clearly higher.

Factors of Production

Economists talk about the four factors of production: land, labor, capital, and entrepreneurship. The productivity of the first three shows up in many ways. We talk about 100 bushels to the acre, 200 parts per shift, or 8000 printing impressions per hour. Entrepreneurship productivity is usually measured by profitability. However, there is no popular or simple way to measure the overall productivity of an operation using all four factors of production in the denominator.

► ***We Get Paid for Our Output.*** It takes most people a quarter of a lifetime to realize that they do not get paid for how hard they work, and some never get the message. In general, we get paid what we are worth to the organization that is paying us. What we are worth depends not on how many hours we work or how tired we are at the end of the day but on our contribution to output—in other words, what we can produce. The only real argument is the ratio of the contribution of worker A compared to worker B. Should secretaries get less than custodians? Should Iacocca (CEO, Chrysler) get less than Smith (CEO, GM)? Should a Russian physician get less than an American one? These are political questions. Not everyone, assuming they are all producing to capacity, gets paid the same amount because the work activity in various jobs is different. There is a hierarchy of work activity, going from a simple thing like putting a screw into an engine, to building an engine, to building a car, to producing cars for a market. Putting the screw into the engine requires less thought, less planning, and less control and less is at stake than producing cars for a market. This is why group vice-presidents earn more than assemblers.

2.3 Measurement and Control

► ***Productivity Is Global.*** The idea of productivity is a global concept because it encompasses many broad concepts over a wide spectrum. It is possible to look at productivity from many perspectives. If all the people who put the screws into the engine are effective and if all the people who put the engines in the cars are effective, and so on up the line, then operational results will be good. Setting the specifications and delineating the motions to put the screw into the hole are straightforward. Pick up the screw in your right hand, place it in the hole, and twist clockwise as far as you can go might be a set of instructions. Measurement of productivity in this case is simple also. How many screws ended up correctly in the engine in an 8-hour shift compared to the normal or standard rate? At the other end of the continuum is the employee who thinks, plans, controls, and takes risk for marketing a line of cars. Setting the specifications and delineating the results are much more difficult for this "worker," as is measurement of control and productivity. It is essential that it be done, but, in fact, it is done infrequently or in a purely subjective fashion.

Management by Objectives

Efforts are made in industry to measure the performance of middle management in a scientific way. The name of one technique is management by objectives (MBO), and it is effective if implemented properly. The theory dictates that we let the individual set her or his own measurable goals in concert with a supervisor. At the end of the period, merit pay is determined on the basis of how near to the objectives or goals the manager comes. However, we have been studying the direct-labor individual for longer and in greater depth than vice-presidents. Because this situation must change if we are to be competitive, it will.

▶ ***The Worker Has Been Extensively Studied.*** The activities of the production worker were first studied by Frederick W. Taylor over 100 years ago and studies continue today. One reason the shop floor worker has been studied for so long is that it was believed, if management could identify and delineate the tasks and motions correctly, productivity would increase. The underlying assumption was that, like the machine he or she attended, the worker was a stupid and insensitive person whose only need was a weekly paycheck. In Taylor's day this may have seemed the case, but it is doubtful if this was ever true. In any event, such thinking by management persisted and still does in some quarters. The culmination may have been the Vega assembly line in 1973, where workers, through scientific management methods, were required to repeat the same task every 36 seconds during an 8-hour shift. The results were predictable; absenteeism, strikes, and sabotage, with less than planned production.

▶ ***Workers Can Be Motivated if Treated Correctly.*** Enlightened management is built on the premise that the worker is motivated by satisfying higher-order needs such as self-worth. The worker will be productive if we can structure the job so that he or she makes a meaningful contribution to the group as a whole and the contribution can be seen by all. This works in principle and is being employed in certain industries with success. For the most part, the high-tech industries, such as computer builders, are not unionized, and there are companies that will not allow stopwatches on the shop floor.

▶ ***Need to Consider People as well as Machines.*** Thus we need to plan, design, measure, and control not only the machine aspects of a CIM plant but the human factors as well. Twenty years ago, if a worker damaged his $5000 Bridgeport, a type of milling machine, because he thought he was being taken advantage of by management, the consequences to productivity were not too severe. Consider the consequences if a worker throws his wrench into the $50,000,000 flexible manufacturing facility. Yet even now, at this high level of technology, studies are being published that point to high levels of fatigue, boredom, and dissatisfaction among the highly skilled and highly trained people who tend and fix these systems. If not to catastrophic failures, at the least fatigue, boredom, and dissatisfaction lead to stress, and stress leads to impaired productivity. So productivity is, again, shown to be a function of a combination of people and machine considerations.

▶ ***Need for Worker to Be Visible.*** We need to be able to measure and control the productivity not of the individual worker or machine but of a grouping of machines and people in the CIM environment. The worker must identify with the group and must be a recognizable entity within the group. This lesson was learned by looking at the experiences of 3M as presented in *In Search of Excellence*.* Their growth record has been attributed to the formation and nurturing of entrepreneurial cells, where each cell is a discrete entity within the organization and individual contribution or lack thereof is instantly visible.

If we shift the emphasis from merit pay based on individual contribution to group contribution, the first thing that will happen is nonproductive members will be eliminated from the group. Sanctions may range from censure to shunning, but they will be imposed either overtly or covertly. The group will find the most efficient way of utilizing its resources to improve productivity.

Return on Investment and Assets

This brings us full circle on the need to measure and control productivity. The dilemma is to measure and control something that has not been defined adequately. Financial people tend to look at two criteria in particular and, being involved with money, they think in terms of dollars. The first is return on investment and the second is return on productive assets. With investment, we buy a machine for $X and it can produce $Y worth of goods in a certain period. Assets include the machine (less depreciation) and other things, such as inventory, plant, land, and accounts receivables. Using the productive assets method, we

*Thomas J. Peters and Robert H. Waterman, Jr., *In Search of Excellence*. New York: Warner Books, Inc.

need to identify which assets are productive to the specific group being measured, which are not, and which are shared by other groups so that they can be prorated. The process of allocation of productive assets usually involves bargaining, negotiation, and decree, all of which are complicated and time consuming, but straightforward. What should be done about people who are considered neither an asset nor a real investment in terms of dollars? We hear personnel and management people say, "Our people are our most important asset." But where do you see the line item, people, on a balance sheet? You don't.

► ***A Way to Get People on the Balance Sheet.*** A simple means of getting people onto the balance sheet does not exist. However, if we agree the success of a CIM operation is dependent on both personnel and machines, it seems reasonable, if we are to measure the productivity of a CIM plant, to incorporate a people factor in the denominator. To ensure dimensional consistency, it seems logical to assign a dollar figure to the "asset" people and, in fact, bring people onto the balance sheet. We should capitalize people like we do machines and continue to expense their maintenance and hygiene factors, such as salary and Blue Cross. This approach, although unorthodox, could lead to a valid measure of productivity. Time will tell if such an approach is accepted.

2.4 Process Evaluation

Multiple Criteria

The lack of a superior method for evaluating the productivity of a CIM facility should in no way hold back the development of the processes by which we make the goods. We can use multiple criteria for determining the effectiveness of each individual process to guide us in instituting corrections to improve the situation. The main objective is to make parts that are available when needed, but not before, parts that function as intended, and parts that consume a mix of labor, capital, and raw materials so as to minimize the cost of consumption of the mix. This is a very tall order, but it can be broken down further into its individual components. It can be thought of as an exercise in linear programming for which we have a series of relationships and constraints on which we wish to operate to arrive at some goal. It differs from a classical linear program in that CIM is an environment. It covers both the relationships and constraints. Even the goals are not fixed. Models could be designed, but they may be obsolete before the design is completed. Furthermore, implementation of changes will require traversing departmental lines. How do we quantify the politics and compromises that will inevitably result?

Establish a Clear Objective

The process must start with a clear objective as to what the part is supposed to do. Since the individual part goes into assemblies that do something that provides some benefit to the customer, the place to start is with the customer, the person who pays. The customer is the piper and calls the tune, whether the average consumer, the military, or someone else. From the specifications obtained from the user, the product design and engineering can begin, provided no engineering and development effort is required, to pin down technological feasibility. Manufacturing engineering and product engineering must take place concurrently. The importance of this step cannot be stressed enough. It is possible to design the world's greatest widget, but if it cannot be fabricated economically there is not much point in going ahead. United States cars are improving, but foreign cars that provide the same benefits are still less costly to fabricate and in some cases are even priced less.

A holistic or global approach to the engineering takes into consideration not only the fabrication of the individual parts but also the assembly. The elimination of fasteners is a worthwhile goal in connecting parts together. The use of fasteners, as opposed to snapping parts together, is highly labor intensive. Not usually taken into consideration in this situation is the cost to the customer of out-of-warranty repair. A welded unit body is cheaper to make than one that is fastened together, and a car so fabricated can be sold for less. However, what looks like a $100 collision on a unitized body may turn out to cost $1200. The result is the consumer does not pay any less for the unit body. In fact, he or she may pay more, because a part of the total cost is transferred from the sales price of the car, which the manufacturer receives, to the increased premium, which the insurance company receives.

Note that the customer tells the manufacturer what he wants and the price he is willing to pay. Product and process design create specifications that meet these requirements or criteria. Manufacturing makes the product conform to specifications that satisfy the required objective. Figure 2.2 shows a model of this concept.

Quality

In conjunction with product design, consideration must be given to both the level of quality to be incorporated into the product and the manufacturing means to achieve this quality level. A popular misconception is that quality must be perfect in order to pass. If this were true, Detroit would equip their cars with stainless-steel mufflers. The necessary quality level is determined by such factors as what the competition offers, what the customer wants, how long the product will be in service, the use to which the product is put in normal and abnormal conditions,

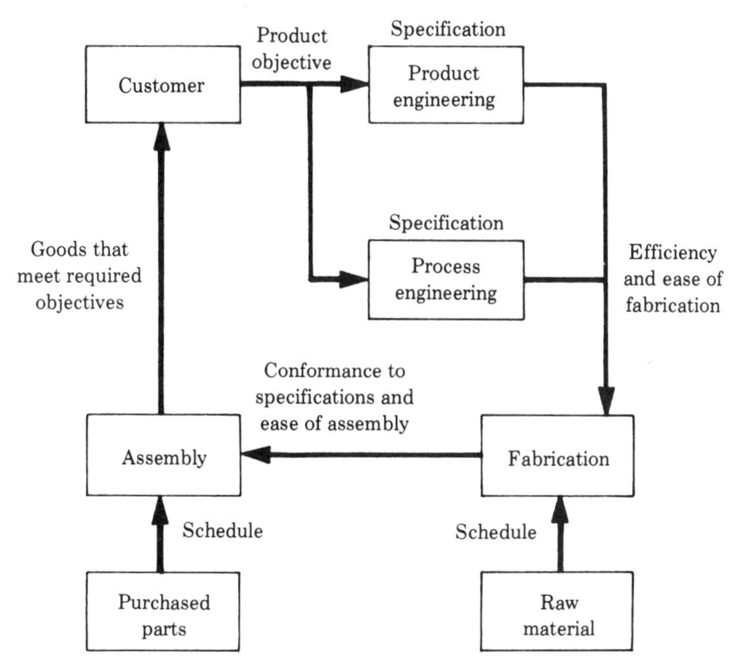

► ***Figure 2.2*** Cycle of Idealized Sale

and the environment in which the product is to be used. Until recently, manufacturers gave little thought to the consequences of customer dissatisfaction from premature product failure, generally failure under warranty. The thinking was that a warranty that covered the cost of repair at no charge to the customer would produce customer satisfaction. Polaroid did a study on this and found that, even though they offered an unconditional money-back guarantee on the return of a bad film pack, only a small percentage of people who experienced product failure actually returned the bad film. What these people did was complain to other people, buy product from the competition, or go to a different format of photography. The policy now is to ship film with no defects and eliminate the need for warranties. Once the decision has been made as to the quality level required and the manufacturing process has been determined, both for fabrication and assembly, attention can be given to the equipment needed to accomplish these objectives.

► ***Stabilize the Process First.*** Two steps are involved at this stage. The first is to determine how the part is going to be made without automation and to stabilize this process so that the part can be made without difficulty. After we know how to fabricate and assemble the parts, then and only then can we look at how to automate the manufacturing process with CIM components. In any fabrication and assembly operation, many parameters need to be studied in order to stabilize the process. Assuming the parts can be made to specification, we need to ensure that the raw materials or stock is on hand, we have the right machines to do the job, the sequences are correct, workers are correctly and adequately trained, production lines are balanced, parts fit together as intended, drawings are correct and reflect reality, vendor activity is coordinated, and more. Stabilization should take place on a pilot line, which is a model of a full-scale production line. Omitting the manufacturing stabilization step has been the most significant cause of problems and dissatisfaction with CIM. All too often we hear CEOs lament, "We put in a factory-of-the-future and expected great results. It has taken us six months to achieve even 50% of what we planned on and we are way over cost estimates. CIM has let us down." If CIM is to achieve results, management must do all the required homework. The Japanese know the kind of product they are going to offer, the market for the product, its quality level, and exactly what steps are required in the manufacture of the product before they automate their factory, and thus they suffer less difficulty in this regard. The place to stabilize the process is before the product gets onto the line.

▶ ***Designing for Parts Assembly.*** As stated earlier, there are two major steps to a manufacturing process. The first, fabrication, is the actual forming or cutting of the metal or other material. The second is assembling the components. Serious problems can develop in both areas if the parts are designed so that they are difficult to fabricate and assemble. The problems can be lessened if design engineers work in concert with the engineers who make the product. Unfortunately, mainly for political reasons, there is usually little communication between design groups and manufacturing groups in U.S. factories. Polaroid, a notable exception, has extensively studied the situation as it relates to cameras. For example, their SX-70 case is now molded so that it snaps together, whereas before the case was held together with screws. This change has saved thousands of work-hours in direct labor, not to mention the savings in not having to inventory screws. Other things that can be done to make assembly easier include no adjustments, no cables, a single power supply, no paint, no stick-on labels or logos, and configuring stampings so they do not hang up in parts feeders. Also, things like frequent engineering changes, adding optional features on an automated line, and not building to schedule will severely damage productivity.

▶ ***Reason to Automate.*** The compelling reason to automate is to produce consistent products at a steady rate. It is fundamental that goods that are not available won't be bought and things that do not work right won't be bought again. One argument for automation is that it reduces direct-labor costs by reducing the amount of direct labor involved in the production of the product. This is true, and one result of total automation is the elimination of all direct labor. The "factory without lights" is a term sometimes applied to this concept. There are two things wrong with this type of thinking. First, factories that have had a very large percentage of direct costs in relation to total costs have long since disappeared from the American scene or are in the final stages of dying out, as is the case with the shoe industry. Direct labor is not the most significant cost, as mentioned before in Chapter 1. Second, what we lose in direct labor we make up for in upgraded indirect-labor talent to a large extent. We need people who can program and maintain the equipment. It then becomes a matter of education and training, which means that value has been added to the employee, who must be so compensated. Experience has shown that increasing productivity, through automation or any other means, eventually leads to increased employment, not less employment. However, during the transition from one technology to another, there may be periods of unemployment in certain sectors. A basic economic principle is that more goods will be consumed at lower prices. More people are employed in the computer industry in 1985 than in 1965 because more people can afford computers. Loss of market share through failure to take advantage of technology is the main cause of sustained and pervasive unemployment.

► ***Islands of Automation.*** An analysis of all the operations in a manufacturing process will reveal which parts of the process can be automated now, which will likely be automated sometime in the future, and which will never be automated. (It is assumed the analysis will be unbiased and consistent, which in reality may be unrealistic, as we may want to tell the boss what he wants to hear.) What this approach has lead to is islands of automation. We find a robot in one place painting doors, a machining center or work cell in another place making shafts, and a drafting design department doing drawings on a computer in a third place. The broader challenge is to tie the islands together under one coherent scheme. This is what a CIM environment is intended to do. Figure 2.3 shows one method of integrating the factory with a host computer. The degree to which the factory becomes automated will be dictated by economics.

► ***Identify Hidden Costs.*** Our decision to automate an island, the first step in CIM, will depend on whether the marginal return on the investment is greater than the marginal cost of the investment. In this context, we are using the terms as an economist would use them, whereby cost includes normal profit as a component. Or, in lay terms, do we get a bigger bang for the buck or not through automation? To

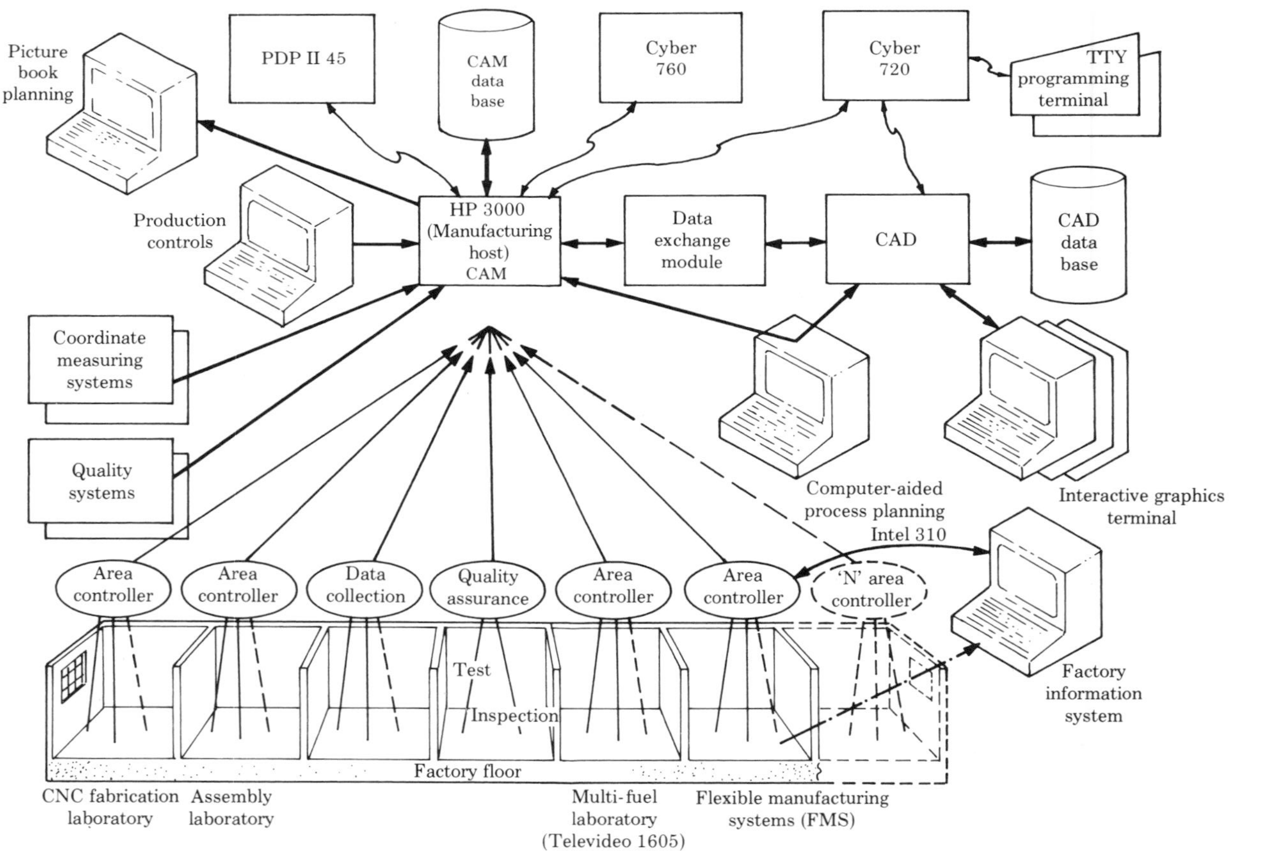

A typical CIM facility uses a host computer to tie the various functions together.

▶ ***Figure 2.3*** CIM Facility (Courtesy of California State University, Fresno, Department of Industrial Technology) determine whether it is to our advantage to automate, we must make sure we have identified all costs, including opportunity costs. Depreciation on the machinery and maintenance costs are straightforward and visible. But what about the cost incurred regaining a customer lost because of poor quality in prior years? How much extra will we have to spend on personnel relations and retraining as a result of the automation step?

2.5 Project Planning

Definition of a Project

Your boss has appointed you captain of the project team responsible for automating the front-panel assembly area. What do you do (besides panic)? Since the implementation of a CIM component has all the characteristics of a project, it is a project and should be handled as such. Fortunately, project planning and management techniques are readily available and you may have been exposed to them already. The key elements of a project are a definite beginning, a definite ending, and certain events and activities that take place between the beginning and end. It is assumed that the project is worthy enough to be funded. Very few projects, and only very simple ones, are made up of a series of sequential steps. Get the money, go to the store, and buy this book is an example of a simple project. Nearly all projects have a multitude of sequential steps, many of which can be done in parallel or concurrently with other sequential steps. The usual challenge is to get from start to finish using the minimum of resources, whether time, money, or personnel. Each step involves an activity of getting to an event, subgoal, or milestone. Tools are available to organize and manipulate the activities and events, and these are to be found in the realm of network analysis and linear programming. Both of these disciplines have been thoroughly computerized, and programs to handle the manipulation are available and inexpensive. Network analysis of projects goes under the name of the critical path method or PERT (see Figure 2.4). Simple projects can be analyzed manually using milestone or Gantt charts. No matter what analysis method is used or whether the computer is employed as an aid, the success of project analysis is solely dependent on whether all the activities have been identified and quantified and whether the events are ordered properly. How long it will take to do this task, how much it will cost, and how many people it will take are activity-type questions that must be answered. What event(s) precedes and follows the event(s) now being considered must be answered as well. Only when these questions have been resolved can the analysis proceed.

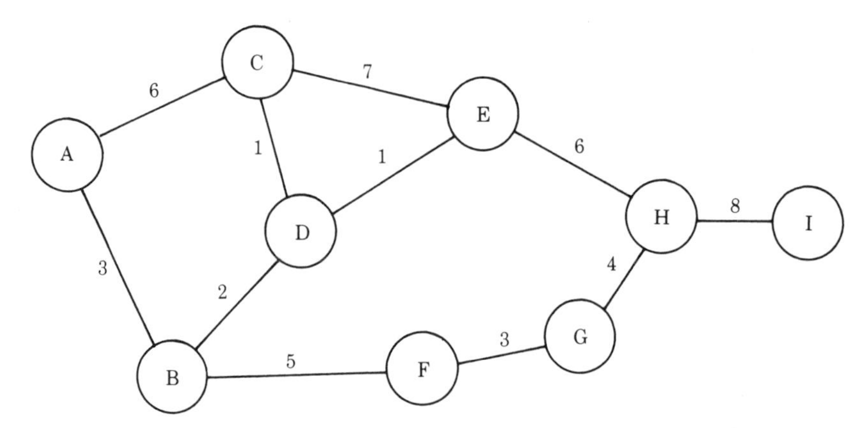

A through I are events; the paths between events are activities and hours to achieve the event. The *critical path* is the longest time path to get from A to I. In this case A, C, E, H, and I, which is the critical path, can be done in 27 hours. The idea is to work on reducing activity times in the critical path to reduce the time to finish the project.

▶ ***Figure 2.4*** A Typical PERT Network

▶ ***Project Management Is Very Difficult.*** Coming up with a project plan will seem like child's play compared to the next step, implementing the plan. The reason has to do with the structure of project management. In a line organization the boss makes the decisions. A good leader will immerse himself or herself in participatory management, be cognizant of and practice MBO, use all the latest behavioral techniques to motivate (not manipulate) the subordinates, get out onto the shop floor to listen to what the workers are saying, and do all the other things a good manager does. The fact of the matter though is, whether the boss is a good manager or not, he or she is still directly responsible for the merit raises and advancement of the subordinate. Notwithstanding "We're just one big happy family here," the boss knows this, as do the subordinate and everyone else in the factory. It is generally considered risky to tell your boss to do something that might be uncomfortable.

▶ ***Projects Use Staff Relationships.*** A project leader's relationship with others is a staff relationship. This is true even with people who hold a lower position in the organization. The key characteristic of a staff relationship is that the staff person's function is advisory. It is presumed the staff person is an expert in his or her field, and so it behooves the organization to listen to the staff person. Some organizations go so far as to make it mandatory that line people follow the advice given to them by staff personnel. Although the project leader is certainly influential in questions of pay and promotion for the people who are reporting to him for direction, the subordinate's line supervisor has the last word. Whatever cooperation the project leader can garner has to come from the

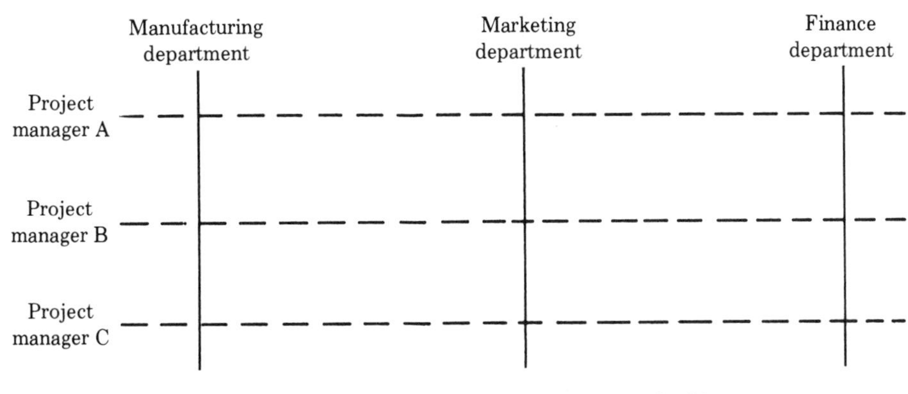

▶ ***Figure 2.5*** Management Grid Structure for Project Management

persuasiveness and salesmanship the leader can muster from within. If everyone in the organization has the same goal, the leader's job is somewhat simpler. Figure 2.5 shows the line and staff relationship in a project organization.

▶ ***All Goals Must Be the Same.*** In a situation where management's objective is to automate the factory, but this is not necessarily the objective of anyone else, the project leader's job becomes several orders of magnitude more difficult. In addition to infinite patience and skill, the project leader will need a great deal of time to achieve the ends. Until very recently, Japanese factory workers wanted to see their factories automated. This accounts, in part, for why they are ahead of us in getting their factories automated. Even greater resistance is likely to come from the first-line supervisors and middle managers. These people will say OK with enthusiasm, but will probably spend their time thinking of ways to subvert or delay implementation. Right or wrong, they believe they will lose power, status, and security. The immediate target is not top management, whom they cannot see, but the project leader, whom they can. We will expand on this theme in Chapter 12.

2.6 Plant–Manufacturing–Industrial Engineering (PMI)

▶ ***The PMI Department Should Keep Things Simple.*** In most places, the group responsible for designing and implementing the automation process under the direction of the project leader comes from the plant–manufacturing–industrial engineering department. In some factories there are three distinct departments; in others, the three functions, if indeed they are separate, are rolled into one. In still others, the functions are all there but are called something else. In this text, this department will be referred to as the PMI department. What they do there, which is what counts, is to figure out where, when, how, and with what they are going to get the product out the door. A purist will tell you that the product design has been frozen and the manufacturing process has been proved in a pilot operation before it gets to the PMI group. The reality of the situation is that product design changes continue to take place long after the plant is in full-scale operation, and you will be fortunate if you get a working model of the part and a complete set of prints. The situation in industry is improving, so take heart. The selection of what machines to install involves many decisions and trade-offs. The best guide is to follow the KISS principle: Keep It Simple, Stupid! Industrial engineers must guard against the temptation to select and install fancy equipment and systems while losing sight of the overall mission.

▶ ***Use Vendor's Knowledge.*** Your best source of information is the equipment maker's representative. Talk with all of them and pick their brains in order to make informed judgments. Keep in mind, however, that this person's pay is directly related to whether you buy something from him or her, so you may have to dig a little to discover all undesirable characteristics of the equipment. Most vendors will sell you equipment that is somewhat overdesigned for the function wanted. This is particularly true of vendors who have been selling for a long time and have a good reputation. A responsible vendor has a reputation and integrity to maintain and, since they are in business for the long pull, they will go to extraordinary means to ensure your success. It will usually pay you to go along with them and not skimp. Think of the overdesign as insurance against shutting the plant down.

What has been said about outside vendors applies equally to equipment designed and fabricated in-house. There are several reasons why a manufacturing company might want to be a machine builder. The best reasons are to ensure that processes remain proprietary and to keep abreast of technology. The worst reason you might hear is because "it is cheaper." This simply is not true. A potential problem, particularly when business is bad, is that the machine-building division may try to force their machines on other manufacturing divisions. The result can be doing make-work. If the designs are inappropriate, corporate resources are used inefficiently.

2.7 Human Factors

▶ ***Human Factors Play a Key Role.*** Even in the factory without lights, machines occasionally need to be worked on or operated by humans. In a typical CIM facility, there are many instances when ergonomics is a significant factor. Some areas are more important than others and as a consequence have been studied more. If you walk into a typical CAD/CAM area, you will find the lights are dim. They are adjusted to exactly the right level to minimize fatigue, and you can be sure the level was arrived at through extensive research. The person on the shop floor is not so lucky. His or her environment is usually not controlled at all or is inherent in the design of the machinery being attended. For want of direction, the machine builder will incorporate those ideas that they think are correct or that are simple to fabricate. Since the productivity in your factory is dependent on these workers, it is incumbent on you to find out what things will assist in ensuring that productivity is not downgraded through fatigue or boredom. You can start with simple things, such as how far the worker has to walk to read a meter. More difficult, but of greater importance, is the answer to the question, "How can I make this job more interesting?"

2.8 Maintenance

▶ ***Good Maintenance Pays Handsome Dividends.*** It has been said that the reason the Germans did so well in World War II was they kept good maintenance records. This may or may not be true in particular, but it is certain that excellence in maintenance will more than pay for its cost in a factory. As automated and integrated systems become more complex and more interdependent, the need for maintenance increases geometrically. In many factories, the maintenance department is to be found in the subbasement next to the boiler, usually in a grubby cage. This situation reflects management's thinking and attitude toward maintenance, and trouble will invariably appear at the worst time. Most people think of maintenance as oil the motors every six months and keep the trash out of the aisles. But it is much more than this. Instruments need to be calibrated and gauges checked regularly, voltage levels must be adjusted in power supplies on schedule, spare parts must be on hand for every conceivable type of failure, and so on. But of greater importance is the need to insist that the equipment be fabricated so that maintenance is simplified. Injection-molding machines are still being sold that look sleek, but require two days to get their covers off so that a simple shaft seal on the hydraulic unit can be replaced. If this machine were tightly coupled to an automated process in a JIT environment, two days to replace a safety seal, which normally takes 10 minutes, would lead to chaos. It happens all the time, unfortunately. You can swear at the machine builder all you want, but if you are the plant engineer in charge, it is your responsibility.

The Japanese attitude toward machine maintenance is that the machine, as received from the builder, is imperfect. It needs to continually be improved so that 3 or 5 years from when it was bought it is a much better machine than when it was new. Americans tend to live with their problems or buy Japanese machine tools when theirs wear out.

2.9 Summary

Productivity attempts to compare the output of an effort with the input into the effort. The problem is that there are many definitions of productivity. In a CIM facility, it is meaningless to talk about productivity as the value of goods produced per direct-labor hour worked, the standard government measure, because direct labor is now a very small part of the cost of most manufacturing operations. Even though we have trouble defining productivity, we still need to incorporate good management control and practices.

The most important management concept is proper motivation of the employee. The employee will be motivated if he or she is treated correctly. It is incumbent on management that the employee be made visible and be recognized for achievements. Participation both as individuals and in groups is needed. Other management concepts to be implemented are a clear statement of objectives and a thorough understanding of the manufacturing processes involved.

The goal of automation is to make quality parts. Programmable automation should be implemented to make the employee's job easier and not simply to reduce the direct-labor cost. Usually, it is the responsibility of the industrial technologist to effect the implementation.

2.10 Exercises

1. What is the classical input as far as productivity is concerned? Name six other kinds of productivity inputs.
2. Why is output per labor hours of direct labor not applicable to a CIM facility?
3. What are the four economic factors of production?
4. Do we get paid for how hard we work? Explain.
5. Describe management by objectives.

6. What was the traditional underlying assumption about a worker and his or her motivation? Why is this incorrect?
7. With respect to people, what do we need to do in order to get a CIM factory working properly?
8. How are we going to shift the merit-pay structure to enhance productivity?
9. What is the first thing we need to do to begin the process of incorporating CIM? Who do we start with?
10. Why must manufacturing engineering and product engineering take place concurrently?
11. From the customer's point of view, is he or she necessarily getting a good deal by buying goods that have been fabricated by the least expensive methods?
12. What factors go into the determination of the level of quality?
13. What do we need to do before we can look at automating a manufacturing process?
14. Where is the best place to stabilize the process?
15. What is the compelling reason to automate?
16. What kind of people do we need to run our modern factories?
17. How did we get to the situation of having islands of automation?
18. List the characteristics of a project.
19. From a people management point of view, what is the problem with project management?
20. What is the difference between a staff and line relationship?
21. What makes the project leader's job particularly hard?
22. What should be the role of the plant, industrial, or manufacturing engineering department?
23. Is it bad to buy equipment that is slightly overdesigned? Explain.
24. What is the key question you should be asking to systematize human factors?
25. What is the relationship between design and fabrication of equipment and maintenance?
26. What is the Japanese attitude toward machine maintenance?

3

CIM Units: Computers

3.1 Introduction

The technical glue that ties all the components of a CIM operation together is the computer. We must know not only how the computer works but also how to tie computers together and how to make them perform the tasks we would like them to do.

Computer Components

Before digital computers, electronic computers were all analog in operation: 2 volts + 2 volts = 4 volts. Today nearly every computer is digital, and digital is even being used in some applications where analog does a better job. Now, when we talk about a computer, digital is implied. We also think of a machine that has these features: a means to input information, a means to extract information, a means to store information internally, a means to operate on information (add, multiply, etc.), and, in most cases, a means to store information externally. The ability to change the sequences and select various operations at will, or programming, is frequently the difference between a computer and a calculator. But not always. Is a programmable calculator really a micro-micro computer? What about programmable controllers? What about a calculator that has operator-selectable programs and one or two memories? All feature a central processing unit (CPU) that gets information from someplace, does something with it, and puts the information back someplace else. Figure 3.1 shows the components of a typical computer.

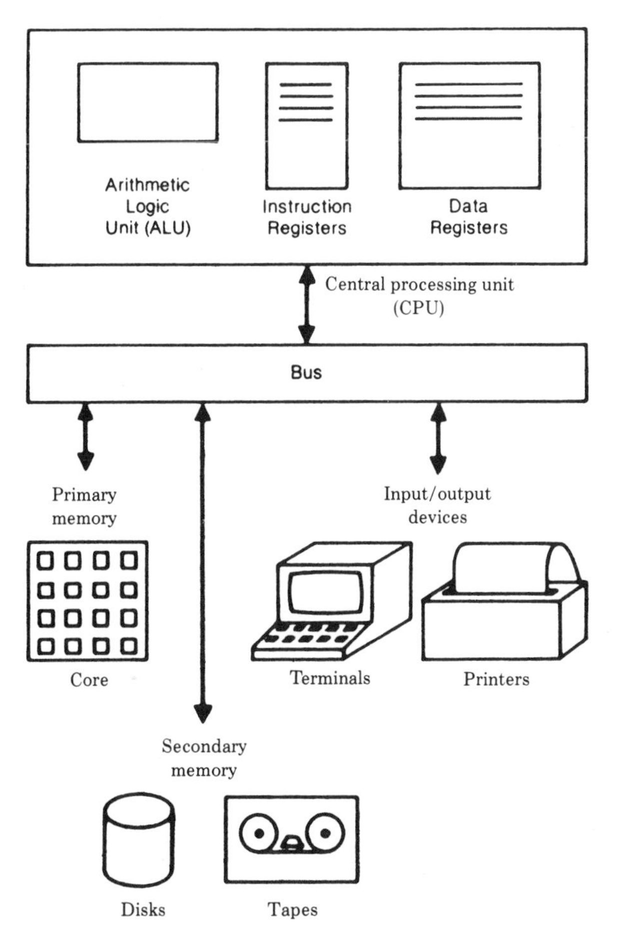

► ***Figure 3.1*** Structure of a Computer System (Harish, *Understanding Data Communications*)

We can think of the process as being similar to a bank of post office boxes. We reach into one box and pull out information. The information is a binary number that is either an instruction, a number representing a real cipher, or an address of another box. The more boxes there are and the faster we move things from box to box the more it costs us.

Rating of Computers

Computers are rated in terms of how large a memory they have and how fast they process information. The most common term for memory size is the byte, which is 8 binary bits. However, depending on which CPU is used, we may have memories that have 16, 32, or 64 bits. We speak of memory in terms of a kilobyte (K), so 1K contains 1028 (2×10^8) addresses or post office boxes. A 64K unit has 64 times 1028, which is really 65,792 but is rounded down to 64,000. Memory is spoken of as RAM (random-access memory), in which we can put anything we want, and ROM (read-only memory), which usually contains programming instructions and constants that we cannot change. (Note, however, that there are special types of ROM that can be reprogrammed externally.) We can also address memory to external storage, such as hard disks, floppy disks, and tapes.

Computer Speed

Speed is determined by a number of factors; the main factor is how fast the synchronous internal clock is ticking. Computers have an internal clock to ensure that one operation does not start before its predecessor has ended. Remember, everything that takes place in a computer takes place in sequential fashion. Each instruction takes a certain number of ticks or pulses of the clock. Two ways to speed up the computer are to combine functions in the instruction set and to speed up the clock rate. Electrons travel around the circuits at speeds somewhat below the speed of light. Thus there are technological limits to how fast you can go.

3.2 Programmable Controllers

► ***PCs Replace Ladder Logic.*** A programmable controller (PC) is another type of computer. An industry friend tells me the term "programmable controller" was used in the early days because there was concern about the reaction of the automotive unions if "computers" were introduced onto the assembly line. These controllers are configured to replace relay ladder logic, mechanical cams, and stepper switches such as may be found on washing machines and dishwashers. The problem with relays and switches is one of reliability, and the problem with cams is that they are usually machine specific and expensive to reconfigure. A typical programmable controller has a simple pushbutton keyboard for programming and data entry and a LED (light-emitting diode) readout or similar monitor on the front. Usually, the panel can be locked to prevent tampering. Sometimes the programming module is portable so that it can serve many processor units, which, once programmed, are reprogrammed only infrequently. The guts of the processor is a small CPU chip such as the 6800 type, a small memory and an output board typically in units of 8 or 16 (e.g., 64) outputs per board. Any kind of output is available, usually on a plug-in module or card.

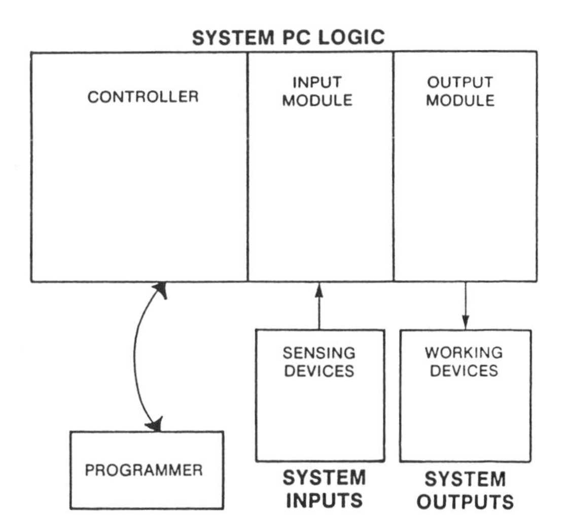

▶ ***Figure 3.2*** Programmable Controller Block Diagram (Courtesy of Digital Equipment Corporation)

Typical outputs are on/off reed or other isolating switches, SCR (silicon-controlled rectifier), voltage from 0 to 10, −10 to +10, and others, current from 4 to 20 and 10 to 50 mA, and computer-compatible logic such as TTL (transistor-to-transistor logic).

▶ ***PCs Are Widely Used.*** Controllers are widely used and inexpensive. They are offered by many firms, such as Gould Modicom, General Electric, and Allan-Bradley. Every manufacturer offers a wide range of features and options. Some of the newer models can be made to interface or converse with larger computers, so the machines they are mounted on can be interrogated or driven by higher-level computers. Programmable controllers are now found on nearly all but the simplest machines, and the market is very large and still growing. Figure 3.2 shows a block diagram of a typical programmable controller.

3.3 Micro/Mini/Maxi Computers

▶ ***Boundaries between Computer Types Are Fuzzy.*** All computers operate in more or less the same way; so when we talk about micro, mini, and maxi (mainframe is a more common term than maxi), we are talking about differences in size and capability. An elephant is larger and supposedly has a better memory than a mouse, although both are mammals. Computer sizes and capabilities form a continuum from a personal computer like the Commodore 64, which has 64K of memory and is comparatively slow, to a large IBM 370, which is in the gigabyte memory range and is very fast. Somewhere in between is the minicomputer, like the Vax11 or PDP11. Some recent micros are bigger than some of the minis, and it is hard to tell where one machine class starts and the other ends.

▶ ***Lack of Compatibility Is the Big Problem.*** Inside the computer, in ROM, are machine instructions that tell the computer what to do and how to do it. These functions are combined by the programmer in any sequence that works (some work better than others) to come up with a viable program. Computer programming is a creative process. Since many sequences are used over and over again, provision is usually made to call up a sequence with a code word that causes the entire sequence to be executed. The problem is that every computer model has a different set of operating sequences and instructions, so no two models are compatible. When we have a high-level language (approaching English) such as BASIC, Pascal, or FORTRAN, which uses the same code words to activate the machine program, each machine may arrive at the same destination, but by a different route. Lack of compatibility in all aspects of the computer industry and with all makes and models of machines is estimated to have retarded the implementation of CIM by 10 years or more. In fact, the lack of discipline among computer hardware builders was so bad that it forced GM to develop its own interfacing systems, called manufacturing automation protocol (MAP). This topic will be covered thoroughly in Chapter 9. The industry is starting to learn that each participant does better if computer use is enhanced by making it easy to interface one computer make and model with another.

3.4 Distributed Systems

▶ ***A Distributed System Is a Hierarchy of Computers.*** An industry that has done a better job of making their computers and peripheral equipment compatible is the process controls industry. In this industry, distributed systems is the term used for a hierarchy of computer controls. It is essential in an oil refinery or chemical plant to have communications between unit operations, because each operation is in lockstep with the preceding or following operation. If there is a malfunction (e.g., too much pressure in the reactor), a sequence of operations is required to shut down the unit or take other corrective action. Before computers and distributed systems, each critical instrument or component was attached to a warning or alarm panel and a light would flash to alert the operator to do something. In newer control rooms, the "do something" is usually done by a computer. The lights still flash and the gongs still clang, but usually the operator is able to override the computer. This sometimes leads to dangerous consequences, like the Three Mile Island and Chernobyl situations.

▶ ***Computers Are Faster and More Reliable Than Humans.*** There are two reasons why computer control is popular in this industry. First, the result of a malfunction is usually catastrophic, not only in terms of death and injury, but from the economics of lost production and destruction of equipment. Because things can happen so fast, rapid corrective action is required. By the time a human operator makes a proper diagnosis and remedy, it may be too late. Second, humans make errors more frequently and with less predictability than computers. A process unit, say a distillation column or mixing chamber, may require a number of automatic valves, temperature, pressure, and flow sensors, pumps, and many other pieces of hardware that need to be read, turned on or off, or manipulated. A programmable controller is frequently employed at the site. Usually, the site is remote or inconvenient to the operations center and so provision must be made to be able to interrogate the controller and override it when necessary. Also, we may want to reprogram the on-site controller from the central station. For these reasons, the computing and control equipment must talk the same language. Furthermore, we need to be able to say, in effect, "If everything is OK, local controller, you run the unit. However, if something is wrong or needs to be changed (a different raw material for instance), our main or central computer will take over."

▶ ***Need for Feedback in a Hierarchy.*** Figure 3.3 shows a model of the control module in a typical distributed system. If we have a number

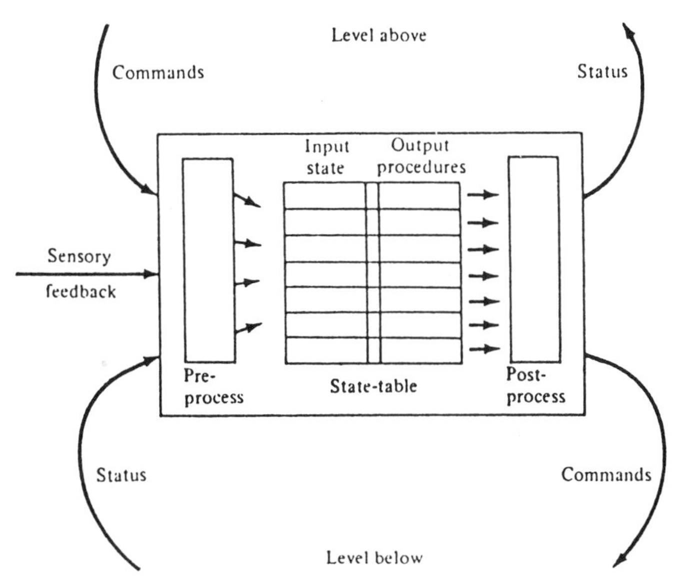

The configuration of each control module, irrespective of the level in the control hierarchy, is diagrammed.

▶ ***Figure 3.3*** Distributed System Control Module (Groover, *Industrial Productivity & Effectiveness*)

of process controllers all tied together in some sort of network, chaos would result if there were not some means of allocating instructions (commands) to the computer module below and information (status) to the module above in the hierarchy. Notice that the module has a sensory feedback input that provides control. Without this feature, the module would be operating in the dark and the system would go out of control. We will go deeper into control systems later.

3.5 Interfaces

► ***Need to Have Everything Match Up.*** When we refer to interface, we mean everything has to be compatible. Just as there is no such thing as "a little bit pregnant," there is no such thing as "not quite compatible." The plugs have to fit the sockets; the wires have to go to the correct places; the voltage, current, and phases need to be uniform; the sequence of pulses needs to be timed right; the formats of the pulses and wave form need to be acceptable; a pattern of pulses from the sender must mean the same thing to the receiver; the sender has to be OK to send and the receiver has to be ready to receive. There are many other electromechanical considerations. Many specific details should be examined before you connect two computers or a computer and some other device (peripheral). When you have come this far, you may be connected physically and electronically, but you may not be able to get a proper response to your message. The computers and peripherals have to speak the same language, and they both need the same operating system.

Computer Language

Language is a more simple concept than the operating system. French, German, and English are languages. So is FORTRAN, Pascal, COBOL, and the other computer languages. I asked a Japanese visitor, "When does your plane leave?" and he responded, "It is, indeed, a beautiful day." We both spoke English, but we were not communicating. The same situation can occur with computer languages. Radio Shack BASIC does not work on a Cyber system. If you try the wrong language or dialect on the computer, you are likely to get a syntax error or some equally helpful message on the screen.

Operating Systems

Think of the operating system as being similar to a large and complex railroad freight yard where they hump the cars. When switches 1, 5, and 9 are set, the car will go to location c and end up in Chicago. When switches 3, 6, and 9 are set, the car will go to location q and end up in Detroit. The operating system directs messages like the signalman directs the trains. Every computer has an operating system and they are all different. The same thing is true with disk drives, which is why you cannot run an Apple formatted floppy disk on a Commodore drive. Consider for a minute what your reaction as a consumer would be if you couldn't play your friend's latest MTV hit on your video player.

Parallel and Serial Interfaces

There are several standard ways of physically and electronically interfacing. These can be broken down into two basic types, serial and parallel. If we want to send an 8-bit byte, for example, we can line the eight 1s and 0s up in a row and send the first bit, then the second, and so on, until all eight have been sent. Sounds simple, but we have to know when the bits are coming and when they have stopped, and we need an internal check to make sure they are correct. Therefore, we send a leading synchronization pulse(s), trailing pulse(s), parity bits (the checking pulse), and sometimes others. Each machine is different; some send 7 bits with parity, some 8, some have 2 sync pulses; there is even parity and odd parity; and so on.

A standard was established, the RS232C, which describes a connector and how it should be connected. However, the standard is not always strictly followed by manufacturers. The specifications on some machines will say RS232C compatible, but upon closer examination you might find some details to be different. A newer standard is the RS422, which is somewhat more "standard." Sounds sort of Orwellian doesn't it? Figure 3.4, which is from Apple's manual, shows the RS232C format. The serial data has the form shown, including a *start bit* at the beginning, an optional *parity bit* after the 5 to 8 data bits, and finally 1 or 2 *stop bits* at the end. This is the *data format* that most RS232C devices use. Notice that, even if we agree to use the RS232C, we still need to negotiate whether there will be a parity bit, what kind and how many data bits we will send, and how many stop bits we will send.

▶ ***A Modem Is Used to Connect Computers by Phone Lines.*** If we must send signals any distance, usually through a phone line, we need a modem (modulator/demodulator). This device takes a tone pulse

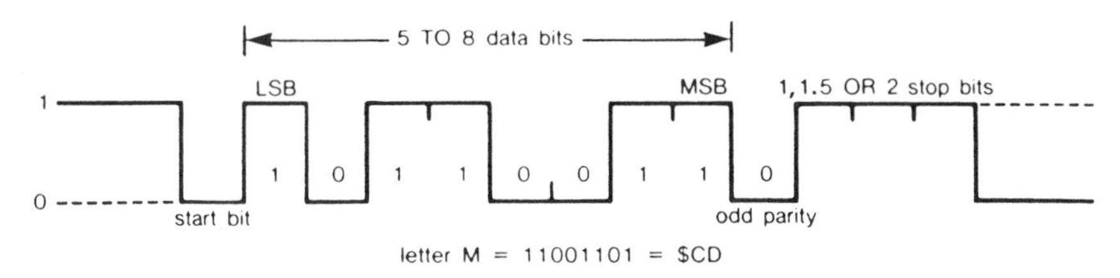

▶ ***Figure 3.4*** RS232C Data Formats (Courtesy of Apple Computer Corp.)

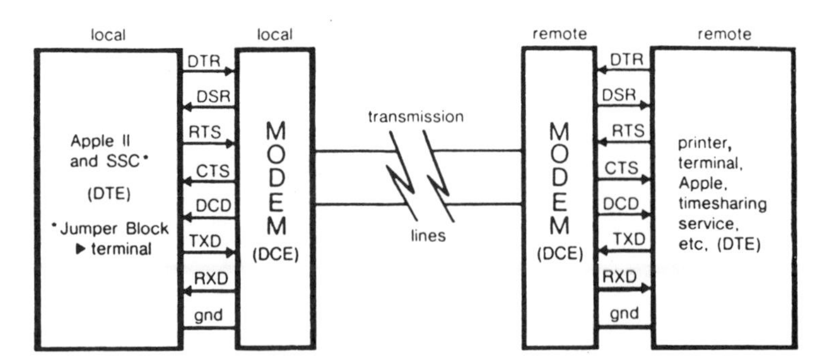

► ***Figure 3.5*** An RS232C Setup with Modems (Courtesy of Apple Computer Corp.)

RS232C Signal	***Abbrev.***	***Similar to***
Data Terminal Ready	DTR	You pick up the phone
Data Set Ready	DSR	The phone is working
Request To Send	RTS	You want to talk
Clear To Send	CTS	The phone has established a connection and the person at the other end is ready to listen
Transmit Data	TxD	You speak into the phone
Not Request To Send	$\overline{\text{RTS}}$	You've finished talking and are ready to listen or to hang up
Not Clear To Send	$\overline{\text{CTS}}$	The phone has sent your words and is ready for your next request to send a message
Not Data Terminal Rdy	$\overline{\text{DTR}}$	You hang up

A. As Interpreted by the Sender

Here are the RS232C signals and how you would interpret them if you were to *answer* a telephone call.

RS232C Signal	***Abbrev.***	***Similar to***
Ring Indicator	RI	The phone rings (optional)
Data Set Ready	DSR	You pick up the phone; it works
Data Carrier Detect	DCD	You hear background noise
Receive Data	RxD	You hear what is said
Not Data Set Ready	$\overline{\text{DSR}}$	The other party has hung up

B. As Interpreted by the Receiver

► ***Figure 3.6*** RS232C Signals (Courtesy of Apple Computer Corp.)

and transforms it into a voltage pulse that the computer can handle. Some of the terminology is quaint as it dates back to the days when teletypes were widely used. Thus a "bell" signal may end up as a beep on your machine. Before transmission can be effected, a number of questions by both the sender and receiver need to be resolved. Is the circuit alive? Are both machines turned on? I am ready to send, are you ready to receive? Many things must be cleared up before the actual data get sent. Figures 3.5 and 3.6, again from the Apple manual, show a modem setup and what the abbreviations mean. Fortunately, when you buy the modem, which is an electronic black box, all you need do is plug it into your computer and the phone jack. The messages sent back and forth are handled by the modem.

► ***Parallel for Local Signals.*** Sending signals 1 bit at a time is slow and cumbersome. Sometimes it is the only way, such as by the phone line, but it requires only one wire and a return. Actually, a two- or three-wire shielded cable is what is used. For short distances (local is the parlance), such as across the room, the parallel mode is preferred. In this mode we send all 8 bits at the same time through 8 wires, one for each bit. The Institute of Electrical and Electronic Engineers (IEEE) has done a good job of setting standards in this area.

3.6 Networks

Bus

An electric bus in a factory is a large copper bar that can be tapped into at various locations to provide power to various machine tools. In a computer, we use buses for handling data and instructions from one part or unit of the computer to another. The signal is put onto the bus at one location and extracted at another, and all parts of the computer are attached to the common bus. If the signal is not picked up by the receiving part, it is retransmitted repeatedly until it is. An alternate method is for the receiving unit to send an "OK to send" message to the transmitting unit. Because every part or unit has a unique address, messages arrive at the proper location as required.

Networks

It is possible to expand this concept to include any number of computers and accessory equipment. Computer people call this a network. Since this is a more recent technology, each manufacturer's network system is different. In general, there are networks designed for computer use exclusively that accommodate digital pulsed information only, and there are broadband networks that handle high-frequency analog video and

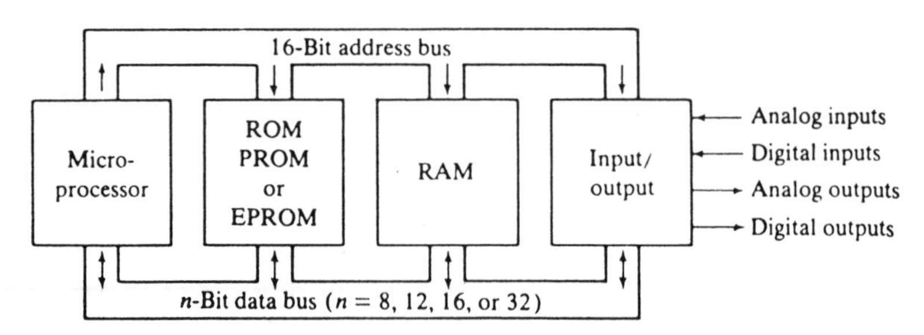

▶ ***Figure 3.7*** A Microcomputer Block Diagram (Bateson, *Introduction to Control System Technology*)

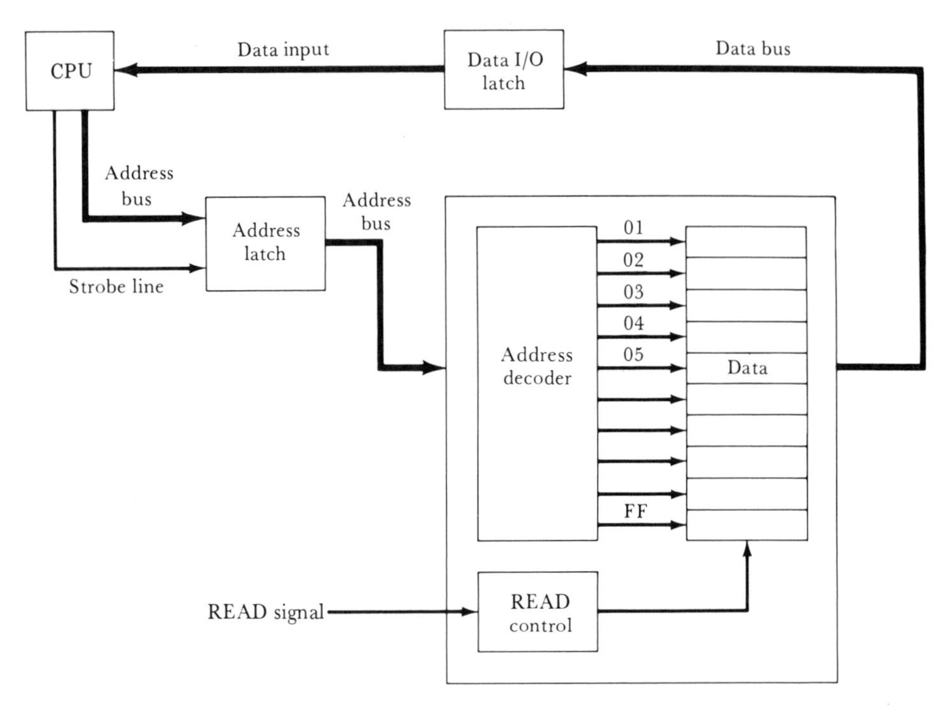

▶ ***Figure 3.8*** Transfer of Data to the Data Bus (Malcolm, *Robotics*)

other signals, as well as the digital signals. A word of caution; not all networks can accommodate all machines. For example, it is not possible for me to interrogate our library computer directly from my terminal due to this situation.

Figure 3.7 is a circuit layout of a typical microcomputer. ROM, PROM, EPROM, and RAM are various kinds of memory chips. ROM (read-only memory) is memory that is rarely changed, if ever. RAM (random-access memory) is constantly changed. ROM is used for things like graphic instructions (e.g., instructions to make the letter A) for the CRT, and RAM is used for data and programming instructions. Figure 3.8 shows the concept of getting data on the data bus. When a read signal is provided, data are placed on the bus from internal and external sources. The processor assigns an address to the processed data, which is then put into memory at the proper location. In this example there are 64 addresses from $00 to $FF (in hex code).

3.7 Peripherals

Just about anything that connects to a computer but is not the CPU or built-in memory can be thought of as a peripheral. Common peripherals include tape and disk drives, screen monitors, keyboards, modems, printers, and a host of other devices that can "plug in." We want to be able to input information into the computer and we can do this in several ways. We can send it the right kind of pulses directly. This is hard for humans but easy for other computers. An alternative is to input using a keyboard terminal, a peripheral. Other common input/output devices are modems and card and paper tape writers and readers. Bar-code readers, voice-recognition devices, and special TV cameras are being developed in many research facilities now. If we want to see what is going on, we need a monitor screen, and if we want a hard copy, we need a printer. If we want to record programs or data (both called files by computer people) for future use we need external storage, such as a tape deck, hard disk drive, floppy disk drive, or one of the newer peripherals like the laser disk. Peripherals come in all price ranges, sizes, shapes, and colors, depending on the user's need, and are available for every type of computer.

3.8 Software

► ***Software Is Crucial.*** Software, or the programs and documentation, is one of the three legs of the information triangle; the computer and storage are the other two. Figure 3.9 shows the information triangle. Remove any leg of the information triangle and it will collapse. Many firms buy the hardware first and later try to fit the software to the hardware. A better method would be to clearly establish the goals, find the software to do the job, and then match the hardware and the rest of the system to the software. In many companies, a central computer

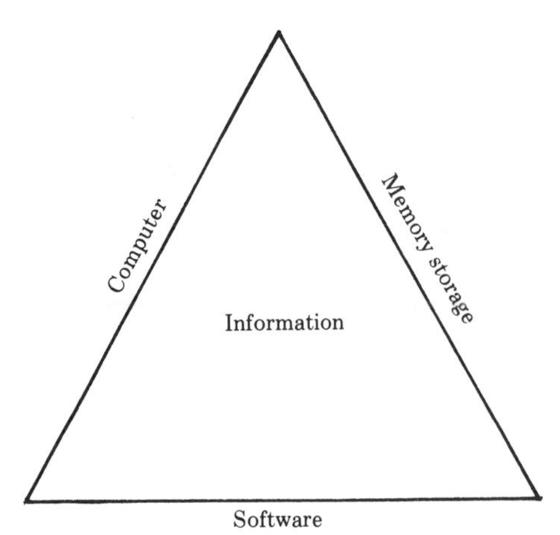

Remove any leg of the information triangle and it will collapse.

▶ ***Figure 3.9*** Information Triangle

group is responsible for finding and/or developing software. However, sometimes these people do not or cannot provide adequate service to the manufacturing units. Traditionally, MIS (management information systems) personnel are "owned" by the financial or accounting groups, so program development for manufacturing sometimes takes a back seat. It becomes a question of politics. In other cases, the MIS group may lack the specific expertise required by manufacturing. A business-trained COBOL programmer might have problems with robotics geometry, for example. Therefore, manufacturing personnel must learn what software is available to do the particular task and be able to write clear and concise specifications as to what is required. As in any specifications, we are concerned with form, fit, and function. We might ask the following questions: "What language is it in?" "How does the operator use it?" "What are the compatibility and interface problems?" "What is the program supposed to do, and will it work in our environment as claimed?"

3.9 Language and Operating Systems

▶ ***Good Language Will Have Good Structure.*** Most schools teach a language first and structure comes along with the language. As engineers, we learn FORTRAN or Pascal. Financial and accounting people learn COBOL. If you take a close look at the higher languages, you will see there are more similarities than differences. Some schools teach structure through pseudo code, and then a machine language or function-specific language is taught later. For example, a do-while or do-until in pseudo code can easily be translated into the language of your choice. However, what is a do-while instruction? What are its characteristics? Where in the program would it be used? What are the consequences of exiting in the middle? If you have taken a language course, you should be able to answer these types of questions. The dilemma in teaching a pseudo code first is that we need hands-on language programming experience to learn structure. However, we need to understand structure to be able to apply it to whatever language is being used in our work environment. Pseudo code, since it cannot be used on any machine, does not lend itself to practical teaching methods.

► ***There Is No Best Language.*** So why not learn the language that is going to be universally used in, say, 5 years? Ask any computer programmer or data processing manager this question and you will receive a different answer from each. Many engineers are learning C because it fits into the Unix operating system developed by Bell Laboratories. However, the situation is by no means resolved. It is an odds-on bet that the language you learn won't be the language used where you will be working.

► ***Need for Uniform Operating System.*** When we speak of operating systems, we are essentially concerned with file nomenclature or how to name files, the relationships between files, and how to manipulate them. In this context, a file can be either a program or data. The file can be in machine language or in a higher-level language such as BASIC. You must know what language the file is in if you wish to use it. Regardless of the file language, you can usually access the file from your terminal. For example, Harry in Kokomo has a data file called "Inventory" that you want to access. The operating system will allow you to do this. Until very recently, accessing Harry's file was impossible because your machine's operating system was different from his. But the computer industry realized that to sell more computers a means to allow computer A to talk with computer B had to be implemented. Bell Laboratories did this with the development of the Unix operating system, and Unix and Unix derivatives are gaining in popularity. Be aware, however, like everything else in the computer industry, there is no standard operating system. Having a uniform operating system in your CIM facility is an absolute requirement. This is why GM and others have been pushing hard for the implementation of the MAP system. You have to be able to get Harry's data file into your computer in order to do anything with it.

The operating system should also allow you to manipulate files in many different ways. For example, you may want to merge or combine Harry's "inventory" with Charlie's "inventory" located in Oshkosh to come up with a total count of product A from both locations. A good operating system such as Unix will allow you to do a wide variety of operations on the files.

3.10 Data Bases and Files

Who Owns the Data

In a CIM facility, each department will have need for data, some of which they will generate and some of which will be generated by other departments. The production planning department needs to know how many units of product A will be sold next month and when they can expect delivery of raw material Q. They can then set production schedules for the individual fabrication units.

The questions of who "owns" the data and who can access what data need to be resolved quickly and reasonably. This is easier said than done for several reasons. A specific department may not have a clear idea what data they need and may overspecify or underspecify the requirements. Similarly, the department may be able to generate information that would be valuable to someone else in the CIM facility but for which they have no direct use. More importantly, whoever holds the data in a CIM facility holds the power. Thus there are political issues that must be resolved. Another question concerns security. Data should be made available on a need-to-know basis and should be accounted for like any other valuable asset.

The temptation is to generate more data than are required in order to take care of contingency situations. It does not take long to clog whatever memory is available with seldom, if ever, used information. Memory, although becoming cheaper, still costs money, so there is a trade-off between memory and the consequences of not having the data in the first place.

▶ ***Data Bases Should Be Planned.*** Certain functions in the handling of data bases must be considered. This activity is described as data-base management. How are the files to be set up so relevant information can be obtained quickly? An alphabetical listing of parts on an assembly might be useful, but a breakdown of parts by part number and subassemblies is essential. We might want to know how many parts on the car start with the letter c, but it is certain we will want to locate P/N Q317A5, which is the carburetor subassembly.

▶ ***Pick Descriptive File Names.*** Naming files can be tricky. A cardinal rule is not to have more than one file with the same name. It is also a good idea to have the file name describe what is in the file. "B23" tells you nothing, but "Carbassy Q317A5" is more informative. Most computers let you use at least 16 numbers and letters for file names, so why not take advantage of it?

What Is DMS?

Data management system (DMS) software comes in all sizes and configurations. A certain amount of space is allocated, called a field, into which you place information up to the capacity of the field. A grouping of fields is a record, and a grouping of records makes a file. You set up the file by stating how large the records and fields are and give them ID numbers and names. Once set up, you can add to, delete from, and manipulate the records and fields to suit your needs. A typical manipulation would be to sort the file in alphabetical sequence. A typical example would be a sort of an address list of vendors with their phone numbers or a sort by subassembly names to identify part numbers.

3.11 Artificial Intelligence

Artificial intelligence (AI) is the newest branch of study in the computer field. It is the study of applying to machines the processes that humans use to make decisions. It is an exciting technology. To understand what AI is and what it seeks to do, a few words about the human brain might be worthwhile.

► ***Left-Side Brain Functions.*** When an author writes a book, his brain is looking at the problem sequentially. He analyzes which words come in what order and combines these words into strings. Sound familiar? Thoughts about what he is going to write may come in random fashion, but through care, planning, and patience he will get the thoughts on paper and then organized into a cogent scheme. The thoughts should be sufficiently ordered and well enough expressed that people can understand them. A specific area in the brain does these kinds of things, and it is on the left side. Sequential functions like writing are called left-side functions.

► ***Right-Side Brain Functions.*** If a car swerves at you while driving down the highway, you do not consciously analyze the situation in a sequential fashion. If you did, you would be dead, because by the time you figured out the situation it would be too late. Something has taken over the thought process and you respond holistically to the situation. A lot of things happen in parallel and "intuitively" so that you get out of the way. Decisions of this kind take place in the right side of the brain. Some people can draw freehand very well. They look at an object and their hand is directed so that the picture looks good. Drawing is considered to be a right-side activity. The ability to take in a lot of information and come up with the right answer quickly is a trait we may envy in other people. This knack to see the "big picture" is a right-side activity.

▶ ***Decision-Making Processes Are Part of AI.*** Scientists have been studying the brain processes with particular emphasis on decision making. Some decisions are made using a set of prescribed rules and some are arrived at through network analysis. Others are arrived at through a trial-and-error procedure. To apply AI, it is necessary to understand what these processes are. Quite a few have been identified, but the surface has only been scratched.

Once a process has been identified, the next problem is to program the process so that a computer can handle it. However, the average computer is a sequential device, and no matter how fast it is, it cannot process the information quickly enough. The solution is to emulate the right side of the brain and process the information in parallel. There is a great deal of activity both in industry and universities to build parallel-processing computers. General Electric and others have developed commercial units, and prices, although still high, are coming down. We are on the road to having both the hardware and the programming for viable AI systems.

CIM Needs AI

Without AI, a CIM facility will not be able to operate at its full potential. Present serial computers are able to process a great deal of a certain kind of information very rapidly, particularly information that comes from a single source. A machine, for example, may call for a tool change and reset in order to make a slightly different part. It is a straightforward matter to program a computer to do this. The program calls for a certain tool to be at a specific location so the machinery can effect the transfer. Motors hum and gears turn and a tool is delivered, but is it the right tool? Someone or something has to look at the tool and make a decision. Since the processing of visual information is a right-side brain function, the first choice would be a human, as humans can make these kinds of decisions quickly. This would be a good solution if humans were reliable, but they aren't. So the desire is to use a machine to visualize the object and make the decision. Vision systems are available to do this, but without parallel processing they are extremely slow.

A popular technique for visual recognition is through silhouette or template comparison or comparison of overall shape. The template approach is shown in Figure 3.10. The visual system must not only be able to see the tool, but it must also have in memory a shape table for every tool that could be used on the machine. If the lighting changes, the tool might be placed by the transfer unit in the wrong place with the wrong attitude; or a tool might be delivered with the same overall shape but with a different configuration cut inside the boundary. In either

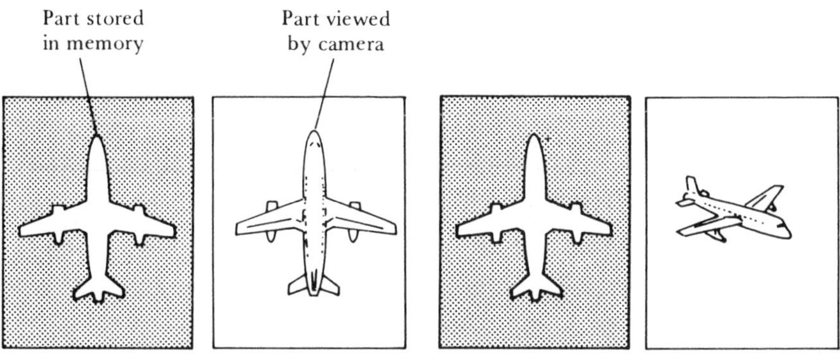

▶ ***Figure 3.10*** Template Approach for Image Recognition (Malcolm, *Robotics*)

case, an error condition will result. And there are many such situations that can cause problems.

We have all kinds of data being generated by different computers and interface devices. Which is relevant and which is not? Additionally, what if some of the data are missing or are in conflict with data from other sources. How do we handle implied information, such as figuring out the meaning from syntax? For example, what is meant by "iron sinks"? Are we describing what happens when iron is put in water or the material from which sinks are made? It depends on how the phrase is used. Programs and systems are starting to appear to handle questions of this sort.

▶ ***AI Will Provide Better Decisions.*** The result of incorporating AI into the CIM environment means better and faster decisions. If a vision system can be designed to pick out the right tool everytime, there won't be a need for a human to be present to make sure the tool and loading are correct. AI techniques are being applied to all types of decisions using whatever data are available. Many of these decisions mimic the kinds of decisions humans make because the human mind makes a good model from which to draw. It is not inconceivable sometime in the future that AI decisions can be made on the basis of models that do not exist in humans. As the data get better, the data bases become broader, and AI programs and machines become more comprehensive and faster, the use of computers to replace people functions will increase in number and level of sophistication. The implications for lower and middle management are profound.

3.12 Summary

The heart of the CIM system is the computer. Here the term computer means any programmable device that has a central processing unit, input and output ports, and the ability to remember data and programming steps. One ability that every computer must have on the factory floor is to be interfaceable with other computers so that all the computers can be integrated.

Although one computer could conceivably be used, all modern systems tie together a group of the computers through a distributed system. Every computer operates at a certain level and passes directions down to lower-level computers and provides information to computers at a higher level. Thus each computer need only be concerned about its own particular job. Control is obtained by providing proper feedback information to ensure that directions are carried out.

To be useful, computers need to interface with peripheral equipment, which collects the information and does the work. Also important are the operating systems, data format, and languages used to instruct the computer. The operating system is a set of internal instructions that directs the activities of all phases of the internal functioning.

A computer system must possess three things in order to be a viable system: (1) the hardware, or the actual equipment itself, (2) software, or a programming language that instructs the computer what to do, and (3) the data base upon which the programs can operate to provide useful results. The three are referred to as the information triangle.

3.13 Exercises

1. How do we rate computers?

2. What is the difference between RAM and ROM?

3. What is a programmable controller?

4. What are the problems associated with conventional relays and cams?

5. What is the major problem with computers?

6. What has been the greatest deterrent to implementation of CIM?

7. What does distributed systems mean?

8. How do we eliminate chaos in a computer network?

9. What is the key element of control?

10. What do we mean by compatibility?

11. What does an operating system do?

12. Describe a serial and a parallel interface.
13. What does a modem do?
14. How is a bus used in a computer?
15. Describe the circuit layout of a typical microcomputer.
16. Name some input and output peripheral devices.
17. Name the three legs of the information triangle.
18. Describe a problem that manufacturing people have with management information systems personnel.
19. Why should the manufacturing person know what software is available?
20. What are the three main characteristics of any specification?
21. What is the dilemma with structure versus language?
22. What is the standard operating system in the microcomputer industry?
23. Who owns the data in a data base and who should own the data in the data base?
24. What is the difference between a field, a record, and a file?
25. What technical development is going to make AI possible?
26. Why are vision systems becoming more and more important?

4

CIM Units: Input/Output

4.1 Introduction

A business office can get by with a computer and memory plus printers, keyboards, monitors, and graphic plotters for making visual presentations. In the manufacturing environment, we need these peripherals plus the capability of interfacing with many different kinds of input and output devices (I/O). Broadly speaking, I/Os do two things. They provide information and they do work. Although computers are digital, the rest of the world is mostly analog; so many devices used in manufacturing convert analog information into digital signals or digital signals into analog power so that work can be done. The voltage and current levels in computers are quite low, so the power-handling ability of a computer is low. Signals are in the milliwatt range. In this chapter we will look at some key elements, starting with output devices and then moving to input devices.

4.2 Electronic, Electromechanical, and Mechanical Devices

► ***Each I/O Has Its Own Address.*** When dealing with digital pulses, we need some way to direct these pulses to do something in the outside world. Therefore, every piece of apparatus has an address. Let's say we want the red light to go on and not the gong, or not both the red light and the gong. We activate a switch circuit that lights the electric light instead of activating a circuit that causes a hammer to hit the gong, an electromechanical device. In a programmable controller, we speak of I/O ports as the addresses and we connect the light to port 1 and the gong to port 2. We can sequence or program the light and gong to do anything we want independently of each other, but concurrently.

Types of I/Os

Devices can be electronic, electrical, electromechanical, or mechanical. A free electron (or a bunch of them) causes a change in a transistor that allows electricity to flow in sufficient quantity to energize a coil of a solenoid, which causes contacts to close, which can carry even more current and higher voltages to turn a motor connected to a hydraulic pump, which forces fluid into a cylinder that moves a steel ingot into a furnace. Sometimes the distinction between electronic and electrical is not clear, but the usual connotation is that electronics is concerned mainly with signal transfer and electricity is concerned mainly with power transfer. Most devices are a combination of electronic, electrical, electromechanical, and mechanical components.

4.3 Bang-Bang and Proportional Control

We can turn things on immediately with all the power that is available or we can feed the power in incremental stages over a period of time in some prescribed manner. An on/off light switch is a mechanism of the former variety, and a dimmer switch (you provide the prescribed manner) is of the latter variety. On/off control is sometimes called bang-bang control.

If we have a set of prescribed sequence (a program), it frequently takes the form of providing high power when high power is required and reducing the power level as the need decreases. This is called proportional control. We find this type of control on things like temperature controllers.

4.4 Motors, Pumps, and Other Power-Train Mechanisms

► ***Speed versus Stiffness.*** Most manufacturing operations require considerable power. Today, all machines are driven by electric motors that are either mechanically or hydraulically coupled to whatever is doing the work, such as a tool or some kind of transport mechanism. A sound foundation in fluid power and electricity provides the basis for the study of the dynamics of the situation.

If we have a cutting head mounted on an XYZ type of robot, we can determine the coordinate path the cutter head must take, and this is straightforward. Since the cutter head has appreciable mass, considerable forces may be required to accelerate and decelerate the head as it goes through the prescribed path. Since materials have elasticity and yield points, we need to be concerned with travel overshoot and equipment breakage due to excessive accelerative forces. To reduce the phenomenon of overshoot with a moving cantilevered arm, we must design the arms so that they have the appropriate rigidity and resonate frequency. The latter has to do with the natural periodic motion of the arm. If well designed, the system is properly mechanically damped and is stable. On the other hand, we want to move the cutter head as fast as we can for productivity reasons. Figure 4.1 shows the concept of the trade-off between speed and stiffness with robot arms as it relates to mechanical damping and stability.

▶ ***Hydraulics Are Used for Heavy Loads.*** Most motors on hydraulic units are of the constant-speed variety, and motion and speed control of the machine are effected by valving. Valve positions are usually set during the machine setup procedure. Because the whole idea of the CIM plant is to have a computer(s) involved in the control, it is sometimes difficult and costly to build in more than rudimentary programming into the hydraulic and pneumatic circuitry. However, for moving heavy loads quickly, hydraulics are very well suited. Transporting engine blocks on an assembly line or moving a heavy welding head into position are applications where hydraulics are used.

Depending on the application, the motors on nonhydraulic machines can be either constant speed or variable speed. Constant-speed motors are much less expensive than variable-speed motors and require fewer components for the drive than variable-speed devices of the same power. Drive refers to the mechanisms and circuitry needed to turn the motor on and off and to modulate the speed.

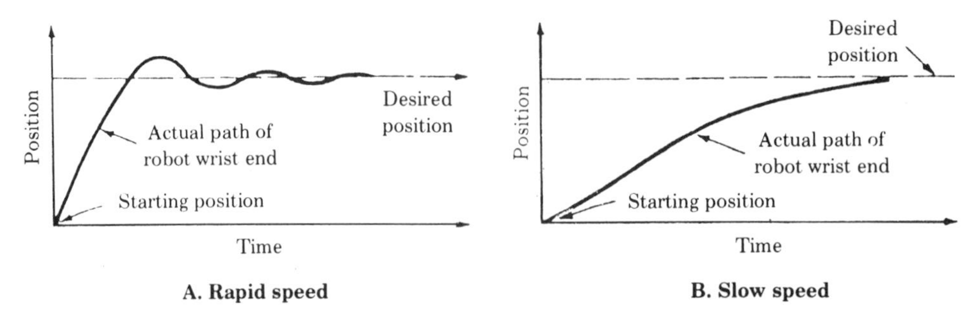

▶ ***Figure 4.1*** Speed of Response and Stability in Robotics (Groover, *Industrial Robotics*)

► ***Fit the Motor to the Available Supply.*** Except in special situations, motors of 1 horsepower (hp) and above operate on three-phase current. Larger motors generally operate at higher voltages so as to draw less amperage, and they thus require less copper in the wires. Even so, a large selection of voltages is usually available for any given motor, because power companies sell electricity with different voltages in different parts of the country. Also, since the available power is ac, proper voltages at the machine site can be obtained through the use of transformers. However, it is less costly to buy a machine with a suitable motor than to fit the machine to the environment with transformers.

► ***Check the Machine Design.*** The pumps, motors, and related power-train devices on conventional machinery are usually sized properly due to the machine maker's experience. In a CIM plant, we are dealing with new machine designs, and so the odds in favor of a malfunction due to improper sizing are increased. Since the machines are significantly more complex, in some cases larger, and usually closely tied to other machines, the capital investment in a CIM plant is a very significant part of the cost of production. Should a malfunction cause the plant to shut down, the economic consequences can be astronomical. The CIM manager may have recourse against the machine builder, but the chances of a prompt and equitable settlement are almost nil. It is better to make sure the design is right before the start button is pushed.

4.5 AC/DC Drives

► ***AC Motors Are More Widely Used.*** Most machines use ac induction motors that achieve a constant operating rpm depending on the load. The principle of an ac induction motor is shown in Figure 4.2, and how it is applied is shown in Figure 4.3 in a three-phase situation. In essence, the induced current created in a wire causes the wire to generate a magnetic field that interacts with the field of the magnetic coil that is providing the inductive field. This interaction creates a force vector in the wire. The mechanisms for turning ac induction motors on and off, called starters, usually employ a low-voltage, single-phase, relay coil that activates the three high-voltage contactors. Additional circuitry may be employed for proper start-up sequencing and thermal and/or overload protection. But, basically, there is an on circuit and an off circuit. It is simple to control this device with a computer or controller. All that is required is an addressable switch.

► ***DC Is Used for Variable Speeds.*** We can obtain variable speeds with either dc or ac drives and the matching dc or ac motor. DC motors are discussed first because the large majority of variable-speed devices are dc. A simplified model of a dc motor schematic is shown in Figure 4.4. Note

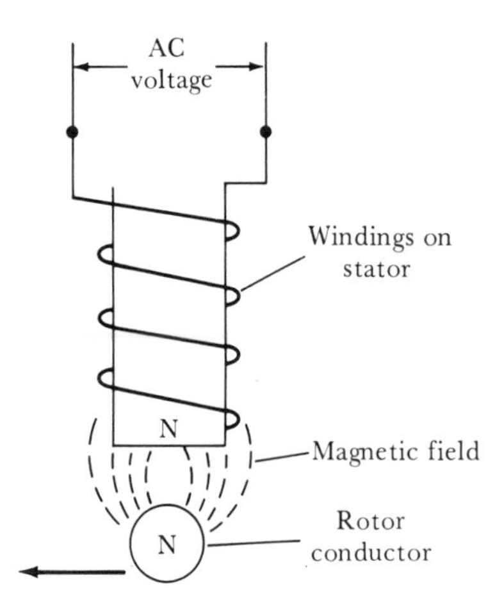

▶ ***Figure 4.2*** Basic Components of an AC Induction Motor (Malcolm, *Robotics*)

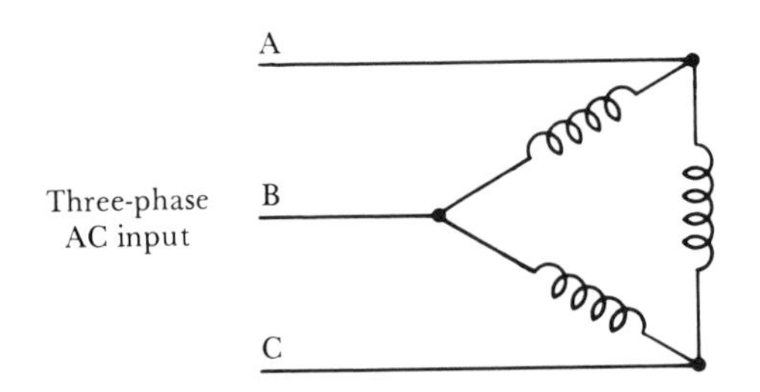

A. Delta connection for stator

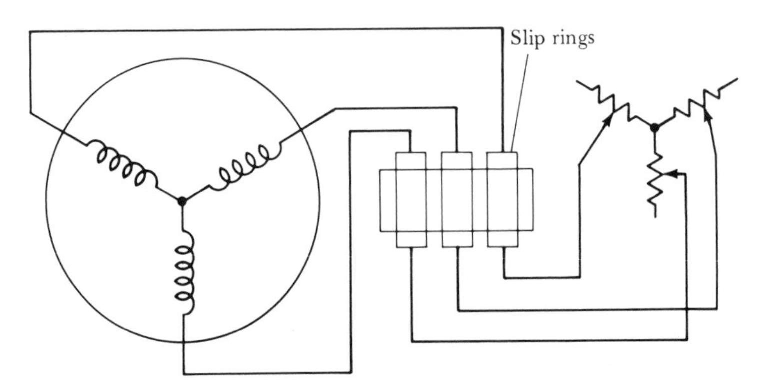

B. Wye connection for rotor

▶ ***Figure 4.3*** Schematic Diagrams for a Wire-Wound AC Induction Motor (Malcolm, *Robotics*)

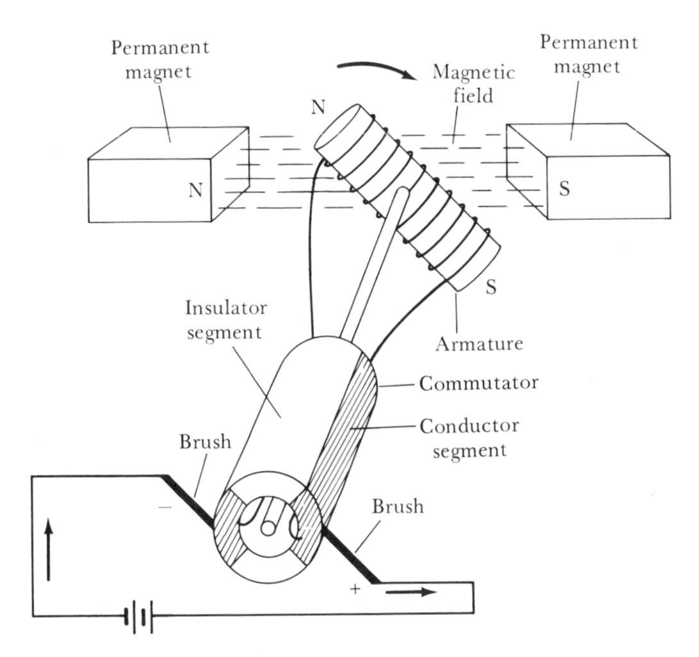

▶ ***Figure 4.4*** Basic Components of a DC Motor (Malcolm, *Robotics*)

that for the machine to work the armature field needs to be reversed. The compelling reason to choose dc is that the load-carrying capability of dc motors over a wide speed range is better than that of ac motors. If we decrease the voltage to slow down a conventional induction motor, power handling drops off quickly. One way to avoid this is to use synchronous motors and vary the frequency. However, it is usually less expensive to buy the dc motor and drive rather than the synchronous ac motor and frequency generator. Motor manufacturers are working on ac systems and the technology is improving rapidly, so we will see more of this type of motor in the future.

▶ ***DC Drives Use All Solid-State Circuitry.*** A typical dc drive takes the ac line voltage and drops it to the desired voltage with solid-state SCR (silicon-controlled rectifier) triac-type circuitry. The current is rectified and may be filtered before it reaches the motor. Figure 4.5 shows a simplified circuit. The speed-adjust resistor is shown as being manually operated. More often, a key feature in most dc drives is a tachometer feedback arrangement that maintains the speed setting

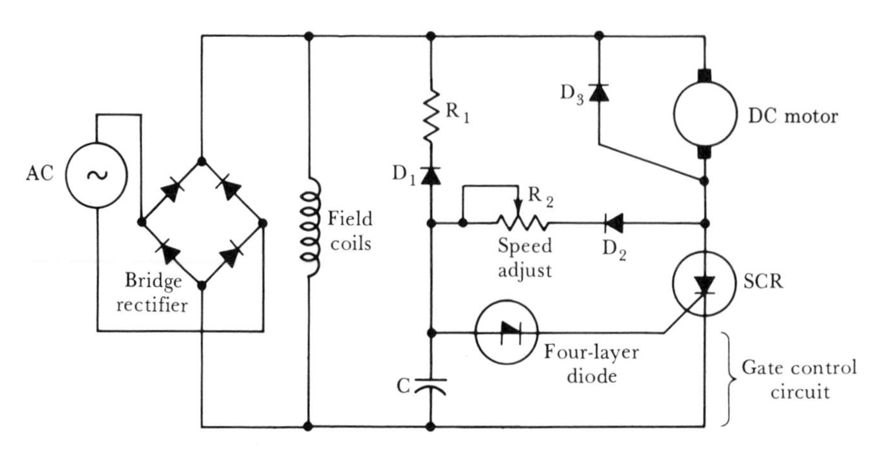

► ***Figure 4.5*** Solid-State Control for a DC Motor (Malcolm, *Robotics*)

as the mechanical load on the motor changes. Most motor-control triac circuits are of the phase angle fire variety. This means that full available voltage flows, but for only part of each cycle. This gives smoother control than other methods, but is more prone to noise and transient problems as the signal source to fire the triac comes from the line. The control knob is attached to a variable resistance that changes the voltage and current levels in the firing circuitry. Thus this circuit is an analog device, and to be computer compatible a digital-to-analog converter must be employed. Fortunately, most dc drive makers have incorporated digital circuitry in the drive or can make provision for it at a reasonable charge.

4.6 Stepping Motors

► ***Steppers Are Used for Positioning.*** A stepping motor operates on the idea that, when given a pulse of power, the shaft will rotate a certain number of degrees and then stop. In the smaller sizes, some are pulse driven by the sine wave on an ac line and can reverse direction by changes in the phase angle between the voltage and current through RC networks. They are very handy, simple to drive, and work well as long as the shaft torque rating is not exceeded.

Other stepping motors are triac controlled, where the speed of motion can be changed by changing the firing rate. Precision is governed by how constant the loading is as the shaft moves, since the motor is an induction device. Because the motor is used as an open-loop controller, you cannot assume that because you gave it the correct number of pulses it actually arrived at its destination. It will not get there if the torque limit is exceeded due to mechanical friction in the machine to which it is attached.

Figure 4.6 shows the sequence of operation of a typical stepping motor. Note that the stepper is operated by turning the fields on and off using switches. The north and south rotor poles tend to align themselves with the opposite pole of the stator, which is controlled by the switching sequence. In practice, the switches are electronic and, instead of two poles in the armature, a commercial unit will have many poles, 50 or even 100. The actual circuitry can become very complicated.

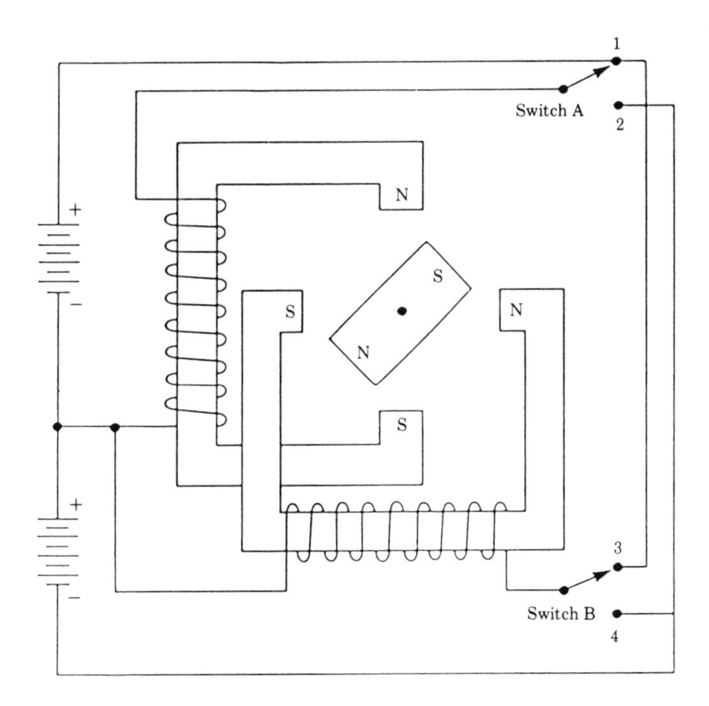

Step No.	Clockwise rotation Switch A	Switch B
1	1	3
2	2	3
3	2	4
4	1	4
1	1	3

Same switch positions: armature has made one complete revolution

Shown are a schematic diagram and switch sequence (clockwise rotation) for a four-pole stepping motor.

▶ ***Figure 4.6*** Stepping Motor Circuitry (Hoekstra, *Robotics and Automated Systems*)

4.7 Encoders

► ***Encoders Provide Feedback Control.*** Encoders are digital devices attached to a machine to provide feedback to either the operator or the computer. They are closely tied in with stepper motors in use and both are used extensively on robots (see Chapter 5). Encoders come in two varieties, incremental and absolute.

Most encoders are optical encoders. If we take a disk, punch holes around the rim, and place it on a shaft, light will pass through the holes but will be blocked by the spaces. As the shaft is rotated, we can pick up the light flashes or pulses with a photodetector to provide a digital on/off signal. Figure 4.7 shows this concept. Since we know how many holes are on the rim, we can count the pulses to see how far the shaft has turned. So many pulses in a certain period of time give us the rate or speed and thus we have a tachometer. This is one type of incremental encoder. The more holes we have in the rim, the more pulses per revolution and the greater the resolution. Most such encoders use a glass disk and the "holes" are placed there photographically as opaque and transparent bars. One thousand lines for a 2-inch disk is common, and the larger the disk is, the greater the number of lines that can be incorporated.

There are two ways to determine the direction of rotation. To do this, we need two channels. We can place a duplicate band beside the first and read it with another optical sensor. For this to work, we have to shift one band relative to the other so that the edge of the line on one is one-quarter of a space or 90 degrees away from the edge of the line on the other. In one direction, the signal senses the two line edges as a 90-degree shift, but in the other direction the shift is 270 degrees. We can do the same thing if we place an identical sensor next to the first one but offset it by 90 degrees. Here, we need only a single band, but we must make it somewhat wider. Both methods are used. If we add still another band with only one line, we can tell when we have reached the starting position. This is handy in case of a power failure that causes us to lose track of the counts. We need to rotate the rotor until we find the start position, however.

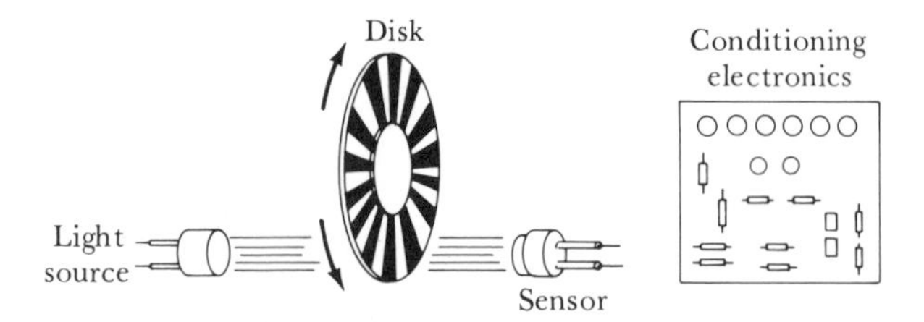

► ***Figure 4.7*** Components of an Optical Shaft Encoder (Malcolm, *Robotics*)

▶ ***Absolute Encoders Do Not Get Lost.*** The way around the problem of loss of position is to employ a series of concentric bands. The first band has one opaque segment and an equal-arc transparent segment. The second band has two sets of equally spaced opaque and transparent segments. The next band has four, and so on. Each band is called a bit, and every position of the shaft has a unique configuration of "on" and "off" band segments. This is an absolute encoder. A 10-bit encoder has 1024 lines on its outermost ring. It also has 10 separate photoelectric circuits and is therefore a parallel device. A 26-bit encoder is used for star tracking and is extremely expensive. Ten to twelve bits is usual for industrial applications. Figure 4.8 shows a 10-bit absolute encoder.

The lines do not have to be on a disk and linear encoders are available. They are usually machine specific and are not as widely used as the rotating variety. They operate the same way as a shaft encoder.

4.8 Servos and Synchros

Servos or servomotors and synchros are analog devices and were precursors for encoders because they were used to provide positional feedback information. Versions called torque converters or torquers are used to transmit power. Synchros are not too widely used in industry, but they are still widely used by the military and in aircraft. In fact, until quite recently the Navy would not use digital encoders on board ships because of lack of confidence in the reliability of encoders under combat conditions.

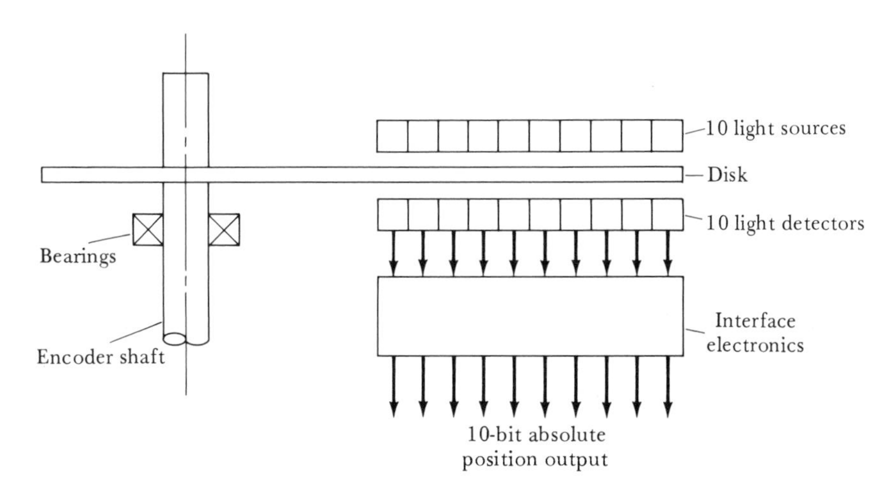

▶ ***Figure 4.8*** Absolute Encoder for Producing a 10 Bit Output Signal (Malcolm, *Robotics*)

▶ ***Synchros Are Analog Position Devices.*** A synchro is an ac device that has its three fields excited by a voltage, which in turn induces a voltage onto its rotor. The three fields are offset mechanically and electrically by 120 degrees. If the rotor is turned, current proportional to the rotor shaft angle and the field flows in the armature. If a similar synchro device is connected to the first synchro, field to field and armature to armature, a turn of one shaft will cause a like turn in the other. A 10-degree shaft rotation to the right in the first synchro results in a 10-degree shaft rotation to the right in the second. But we can electronically simulate the synchro so that we do not actually need a matched pair. We can take the signal from a synchro and read out the position directly or can digitize the information for computer interface.

Synchros come in many sizes, styles, and functions. They are very handy devices and may sometime be rediscovered by industry. Figure 4.9 shows a typical synchro/servo configuration and how the fields are connected. In the diagram, the synchros are shown as a control transmitter and a control transformer, two specific types of synchros. This setup might be used to drive a ship's rudder to the proper position and to verify that it actually arrived at the intended location.

▶ ***Many Things Are Called Servos.*** The term servo is one of those fuzzy terms that has several different connotations. A servo system

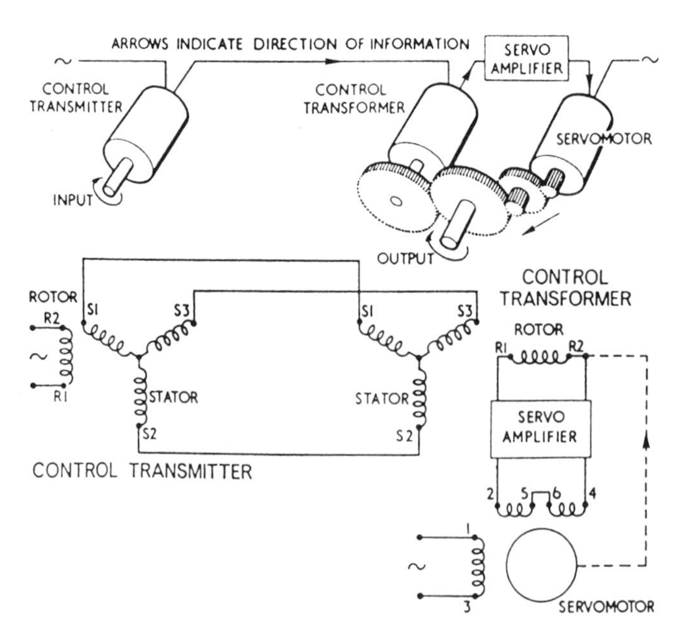

▶ ***Figure 4.9*** Simple AC Remote Position Servo System (*Synchro Engineering Handbook*, Muirhead Ltd)

implies that a feedback signal is used to control the process. Feedback loops are usually for either position or velocity, or both. Position is frequently measured with an encoder or a synchro, and velocity is frequently measured with a tachometer or tach generator. A tach generator is a small dc generator with voltage output proportional to speed. The dc feedback signal is amplified through a servo amplifier, and this amplified signal is used to control the motor. Frequently, but not always, a servo motor has either a tach generator winding and/or a servo control winding in the stator portion of the motor. The servo control signal is 90 degrees out of phase with the field. Voltage changes in the control winding cause the motor to go faster or slower. When the voltages in the two windings are equal, the motor stops. Sometimes the servo (feedback loop) control is an integral part of the motor and sometimes the components are coupled to the shaft externally. The term servo is sometimes used to describe the specific motor package just described, and sometimes it is used to describe any system that employs the feedback principle.

4.9 A/D and D/A Converters

▶ ***Converters Are Usually Supplied by the Machine Builder.*** Nearly all sensors and motors are analog devices. For analog devices to interact with computers, it is necessary to put the signal through an A/D (analog to digital) or D/A (digital to analog) converter. Many instruments now come with a built-in A/D converter, and some specialized power drives are available with digital input. A/D and D/A converters are widely available, come in many configurations, and their prices vary depending on the resolution desired. Typically, you might need an A/D

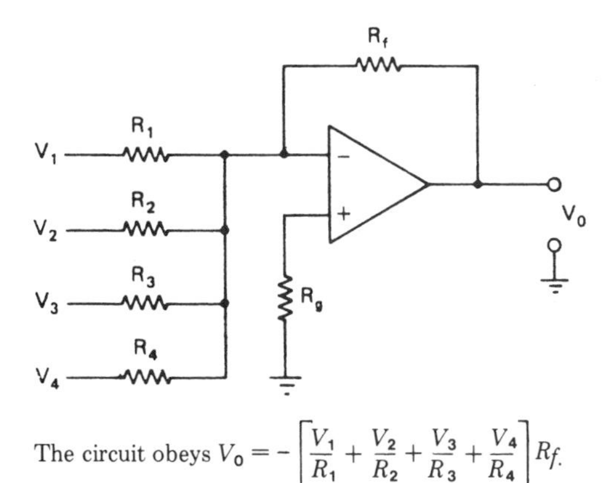

The circuit obeys $V_0 = -\left[\frac{V_1}{R_1} + \frac{V_2}{R_2} + \frac{V_3}{R_3} + \frac{V_4}{R_4}\right] R_f.$

▶ ***Figure 4.10*** Digital-to-Analog Converter (Snyder, *Industrial Robots*)

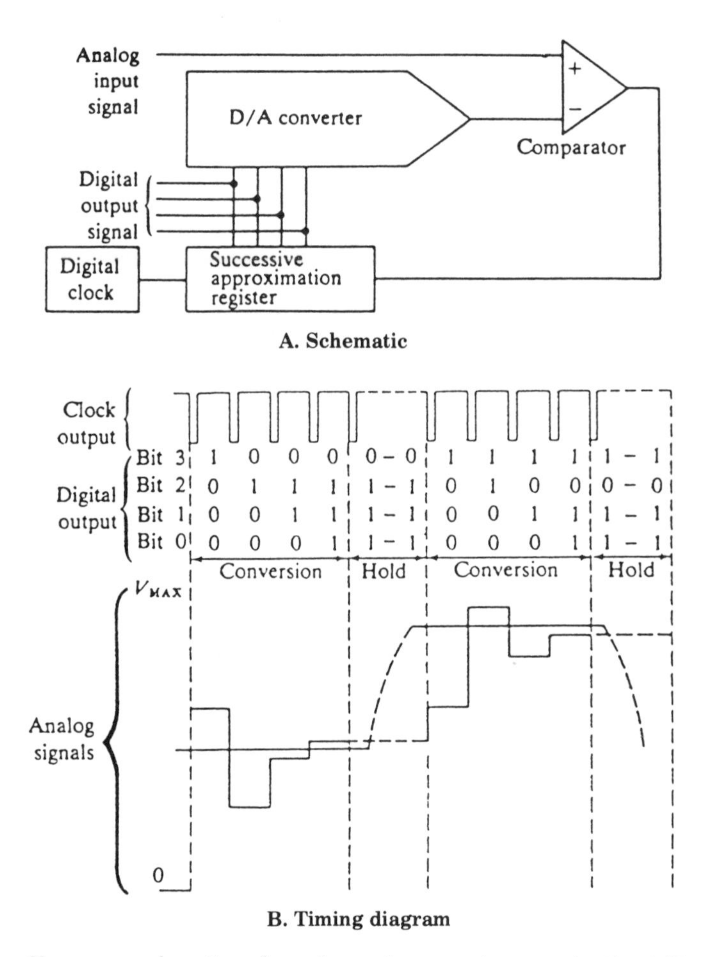

Shown are a schematic and waveforms of a successive approximation A/D converter.

▶ ***Figure 4.11*** Typical A/D Converter (Bateson, *Introduction to Control Systems*)

converter with a 0- to 10-volt input and TTL compatible output with 8 bits. Also available are frequency to digital, syncro to digital, and many other specialized configurations.

The D/A converter is very simple in that voltage signals on binary lines are summed through an op amp (operational amplifier). Figure 4.10 shows how this is done. The A/D converter is somewhat more complex, and there are many schemes for effecting the conversion. One such is shown in Figure 4.11. Digital signals are successively applied until the output of the internal D/A signal is equal to the input signal.

4.10 Visual Sensors

▶ ***A Near Necessity for Robots.*** More and more robots and other equipment are being equipped with visual or tactile sensors. There is always mechanical slippage and friction in any machine, which interfere with precise positioning. In most cases we need to know if the robot arm actually arrived where it was supposed to go. A feedback device like an encoder will help us in this case. However, more frequently a part or tool may be presented to a robot by a human or another machine. If that part is not the right part, is not in the right location, or does not have the right attitude, the robot will not be able to pick up the part or may pick it up and insert it somewhere else incorrectly. Dropping a 500-pound engine block on an assembly line can lead to dramatic results. Since the robot or machine center is computer controlled and since we must know if everything is OK, visual, tactile, and other types of sensors have been developed.

One way to obtain digital picture images is to use a TV camera, which produces an analog video signal, and digitize the signal. The output is linear in the sense that the TV signal scans from left to right and top to bottom. Synchronization and resolution problems to develop good digital images at reasonable speed have made this method less attractive than a pure digital approach. This scanning process is shown in Figure 4.12A and B.

▶ ***Newer Sensors Are All Digital.*** We can take a silicon chip and etch onto it a matrix or grid of small photosensitive dots. Each dot or pixel can be read as being high, low, or, in case of high-resolution graphic devices, somewhere in between. Experimental chips are being produced with 2 million pixels per square inch. Commercial equipment such as used in robotics has much lower resolution, with 10,000 pixels per square inch being more common. We then focus an image onto the surface of the chip and scan the output in serial, in parallel, or in a combination of both.

Here we are dealing with a large amount of digital information, which must be processed with powerful computers. Programming is complicated and extensive. Early programs dealt with static pattern recognition by determining the outlines of parts or a silhouette image. If the image matched the image in memory and was in the right spot, everything was OK and the machine could perform its operation.

Advances continue to be made in this field. One object is to develop systems to recognize different parts placed in random attitudes and locations on a moving conveyor. The visual device must have in its memory all views of each part and be able to pick out the part. A similar problem is to pick out the right part from a bin of random parts even if only a small portion of the desired part is visible.

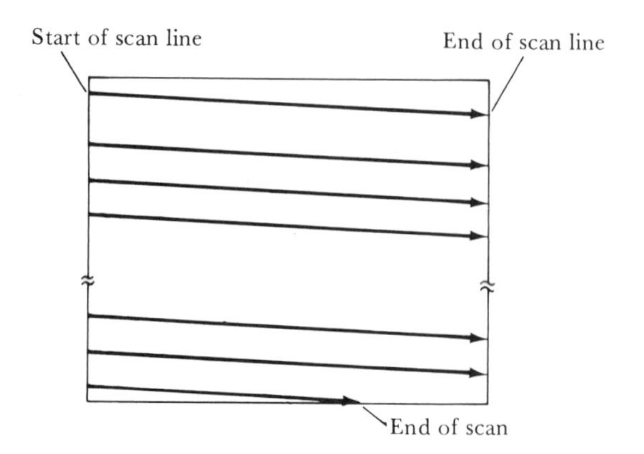

A. Scanning by the electronic beam in the camera

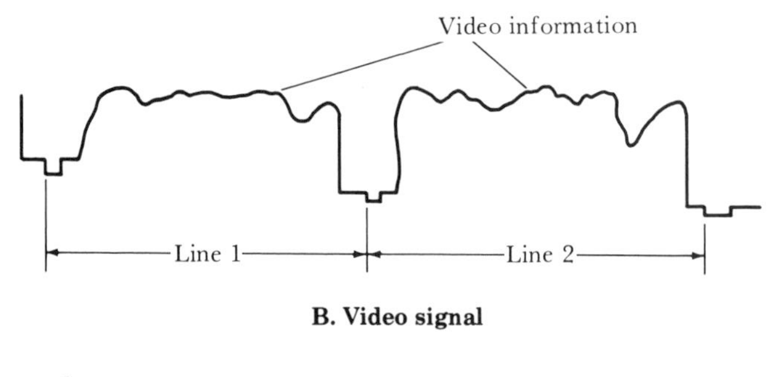

B. Video signal

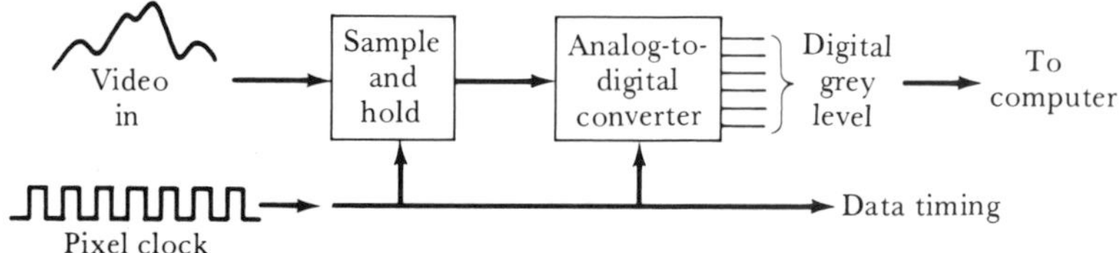

C. Pixel clock for converting a video signal to digital bits

► ***Figure 4.12*** Analog TV Scanning (Malcolm, *Robotics*)

4.11 Tactile Sensors

Strain Gauge and Piezoelectric Devices

A tactile sensor is usually a pressure transducer, whose signal is digitized. Two types of pressure transducers are the strain-gauge and the piezoelectric type. Both have analog outputs and both need additional circuitry to drive them or for readout. The strain gauge works on the principle that the resistance of a wire changes as it is stretched. Strain gauges are commonly used in materials testing and are available in sizes

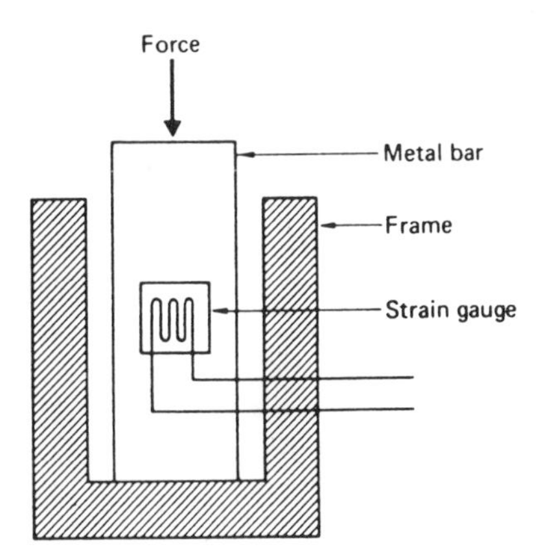

▶ ***Figure 4.13*** Load Cell (Humphries and Sheets, *Industrial Electronics*)

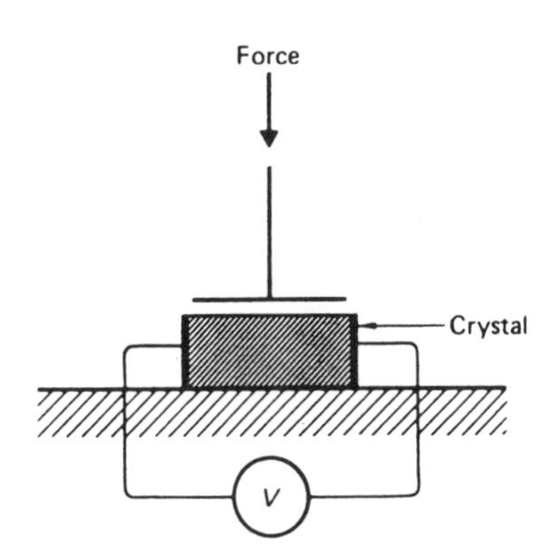

▶ ***Figure 4.14*** Piezoelectric Crystal Generating Voltage (Humphries and Sheets, *Industrial Electronics*)

about 2 millimeters square. Some type of bridge circuit (Wheatstone) is usually employed to read them out. Figure 4.13 shows a typical wire strain gauge used in a load cell. Notice that the wires in the strain gauge are axial to the load.

The quartz crystal is a piezoelectric device. If we squeeze or stretch a quartz crystal, a small voltage will be produced in the process. A precise relationship between pressure and voltage exists and very sensitive transducers are available. From the voltage produced, it is a simple matter to digitize the output for computer interface. This is shown in Figure 4.14. Incidentally, we can apply a voltage and cause the crystals to change dimensionally in a prescribed and exact manner. The

National Bureau of Standards uses this principle for calibrations of measuring instruments.

▶ ***Sensors Keep Getting Smaller.*** Early robots used a limited number of tactile sensors primarily to determine if a part or tool was there or not. The drive to make smaller sensors for greater resolution in a given area has not been as strong as the drive to improve resolution with visual sensors, but it is occurring nevertheless. One goal is to simulate a human finger, which has a very large number of nerve endings and can measure pain as well as pressure. One approach is the whisker sensor shown in Figure 4.15. Manipulators are now available that can handle fragile and nonuniform products like eggs.

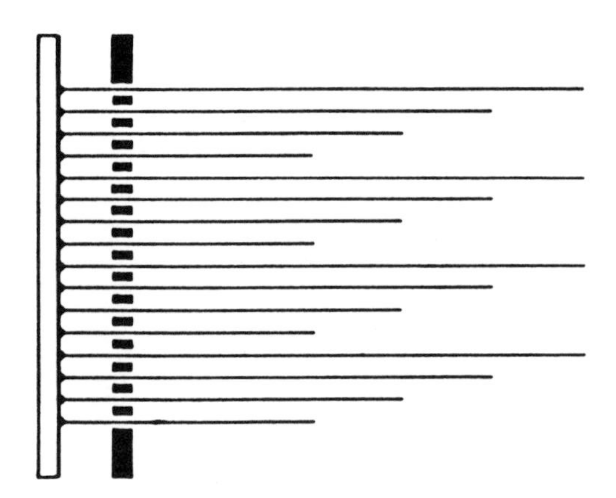

A. Array whisker sensor

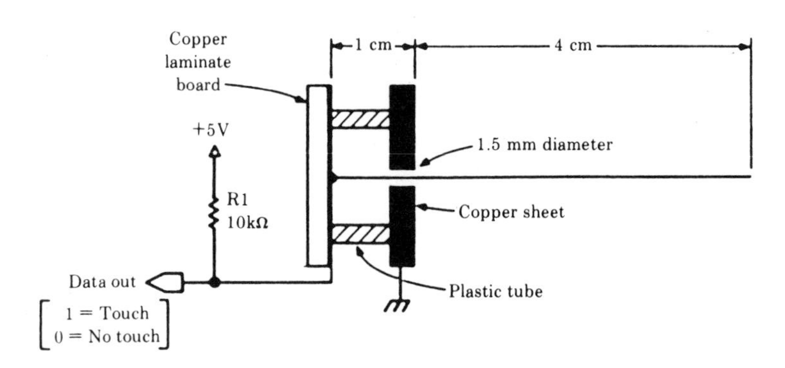

B. Single whisker sensor

▶ ***Figure 4.15*** Cat Whisker Sensor for Robot Sensing (Asfahl, *Robots and Manufacturing Automation* and *Robotics Age*)

4.12 Sonic Sensors

▶ ***Sonic Sensors Come from the Photographic Industry.*** Instant and, more recently, 35-millimeter photography have been revolutionized by the advent of the automatic or instant-focus feature. Polaroid's cameras work on the principle of reflection of a sonic energy burst, a type of sonar. A pulse or burst of energy is emitted by a transponder, travels to the target, and is reflected and picked up by a transducer. The time it takes for the signal to travel to and from the destination is directly related to the distance between the target and source. Polaroid sells a ranging kit for machine builders to incorporate this technology into their robots and other mechanisms. Resolution and precision are comparatively poor, and at present these devices are used mostly as a safety feature. If the arm gets too close to an unplanned object, motion stops before there is a collision. With a tactile sensor, a collision must take place before anything can happen.

4.13 Displacement Sensors

▶ ***Resistance Sensors Are Cheap.*** There are several ways to measure displacement. All are analog and all need an A/D converter in the signal train to interface with the computer. The simplest is the potentiometer (pot) or sliding-wire resistor, which functions when contact with the part is made. For mechanical reasons, the linear pot is more widely used than the rotary variety. These devices are inexpensive and do not require much in the way of electronics. However, resolution and precision leave something to be desired and, as a result, linear pots are used where these factors are not too important.

▶ ***The LVDT Is the Most Popular Transducer.*** The LVDT (linear variable differential transformer), shown in Figure 4.16, measures displacement from a null point by comparing inductances in two identical coils when an iron or ferrite rod moves in and out of the field. As long as the rod is in the center of the fields, the output is 0, but when the rod is displaced, more current flows in one coil than the other. This difference is detected. LVDTs come in various sizes and all have readable resolutions of 1 part in 1000 or so. An LVDT that travels 1 inch can measure to 0.001 inch, whereas an LVDT that travels 1/10 inch can measure to 0.0001 inch, for example. An ac signal is required to drive LVDTs. Note that, since they are a null device, independent means are required for calibration to achieve accuracy. They are used extensively as deviation gauges.

Both strain gauges and piezoelectric devices are displacement transducers in the sense that they require a change in position to

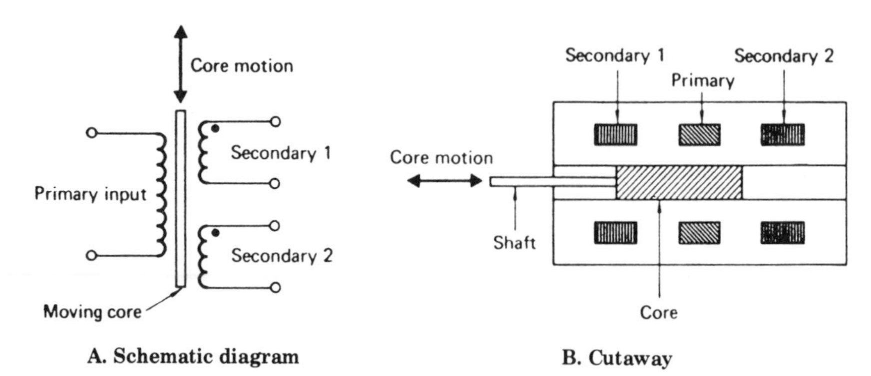

► ***Figure 4.16*** LVDT as Indicator of Linear Displacement (Humphries and Sheets, *Industrial Electronics*)

function. However, since a minute displacement will give a very large output, these devices are not frequently used for displacement measurements except in tensile testing of materials.

► ***Noncontact Gauges Are Used in Critical Applications.*** Sometimes we want to measure displacement without touching the part. One way to do this is to measure the change in output of a photosensitive transducer that is looking at an energy source of some sort. Optical gauges measure either a change in the incident energy, as in the case of changing silhouette images, or changes in reflection energy due to changes in the angle of reflection. Energy sources can be infrared, visible, ultraviolet, or coherent (laser) light. Optical gauges come in many varieties and styles for various applications. For example, a popular laser gauge used for measuring wire diameter sweeps a beam across the wire, and the sensor registers the relative size of the shadow made by the wire.

The photoelectric eye is widely used in industry. Its major use is to detect when a part is present on a conveyor line, or a similar application. The presence of the part breaks a light beam, which is picked up by a photosensor. This is a bang-bang sort of switch and can be used to activate other functions. When connected to a computer, the signal is used to begin or end specific sequences. It is simple and cheap. Usually, infrared sources and detectors are used so that incandescent light won't affect the output and the system can be used in dusty or smoky environments. Fiber optics devices, which can be thought of as light pipes, are finding use in applications where space is at a premium. Because the signals from fiber-optic devices can be amplified, detectors can be installed in low-light-level areas. The night scopes on spy-television are a sophisticated application of the technology. Industrial uses, to date, are still simple and mundane, however.

Inductance and Capacitance Gauges

Another popular noncontact gauge is the inductance gauge. It measures the inductive field between the sensor and the target, which must have magnetic properties. The closer the sensor head comes to the target the smaller is the air gap and the stronger the signal. An RF signal is required and, because the output is nonlinear with displacement, linearization or curve-fitting capabilities are usually incorporated either before or after digitizing.

A first cousin of the inductance sensor is the capacitance sensor. This measures the change in capacitance between the sensor and a conductive target. The air gap between the sensor and the target is the dielectric and for practical purposes is constant. Again, as the air gap decreases, capacitance increases in a nonlinear fashion. Driving and readout circuitry are similar to the inductance gauge.

Resolution in both is about the same and is on the order of 1 part in 1000. Both are sensor area dependent, in that the larger the sensor area is the farther away from the target the sensor must be. Capacitance gauges have been built that can measure high-speed bearing runout to 1 millionth of an inch; they employ a sensor area the size of a pinhead and a gap or standoff of about 0.001 inch.

The choice of using capacitance or inductance gauges for measuring steel in air is a toss-up. Inductance gauges work under water and in oil, whereas capacitance gauges work with nonmagnetic conductors.

4.14 Physical Sensors

In addition to displacement, we also need to measure other physical properties. Temperature, pressure, and flow are the most common.

Temperature Sensors

Temperature is measured with thermocouples, RTDs (resistance temperature detectors), and thermistors. All produce an analog signal that needs to be digitized for computer interface. Thermocouples work on the principle that a junction of two dissimilar metals will produce an electromotive force (emf) or voltage when heated. Figure 4.17 shows the idea of a thermocouple sensor. The higher the heat is, the higher the voltage output. A large selection of thermocouple (TC) junction materials is available; their use depends on the temperature range and service required. The measuring junction is compared against a reference junction of the same materials immersed in ice water (0° C). In modern instruments, a printed circuit in the gauge is used as the reference and is matched to the junction type selected. The circuitry also linearizes the output. Thermocouples have been in use for more than 50

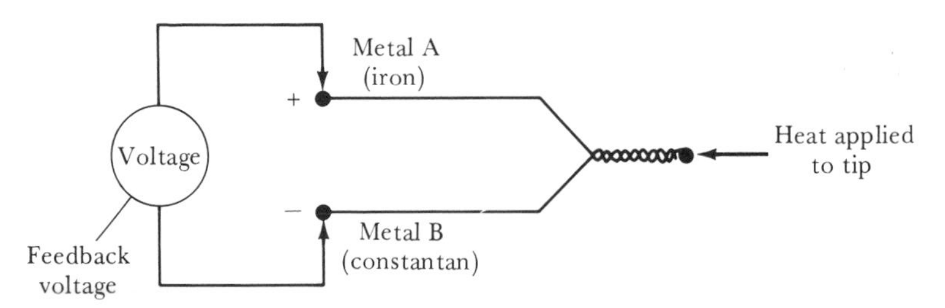

▶ ***Figure 4.17*** Thermocouple Sensor (Malcolm, *Robotics*)

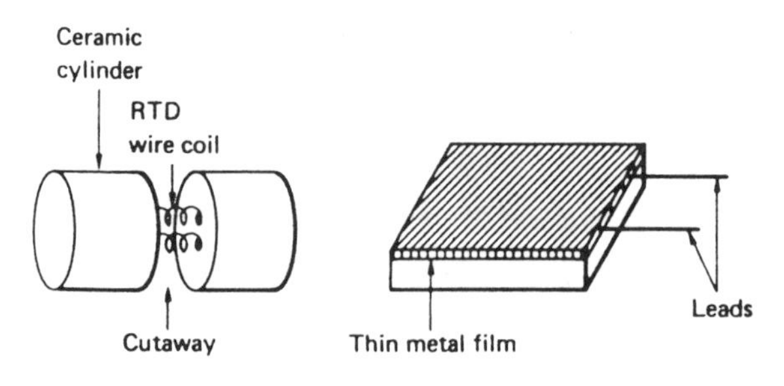

▶ ***Figure 4.18*** Two Popular RTDs (Humphries and Sheets, *Industrial Electronics*)

years and are well understood. They are used extensively at higher temperature ranges, such as found in ovens and furnaces. An often overlooked characteristic of thermocouples is the fact that the lead wires from the TC junction to the gauge box are of the same materials as the TC itself, a special pair of wires in other words. If you change, as opposed to replace, the TC, you need to change not only the printed circuit card but also the lead wires.

Resistance temperature detectors (RTDs) work on the principle that the resistance of a wire changes as the temperature changes. Many materials can be used, but platinum is favored because of its sensitivity, durability, and ability to operate reliably in modest temperature ranges. Platinum wire is not cheap, however. The electronics are simple and straightforward, and the RTD element is available in very small packages. Because of these features, RTDs are found in critical-precision measuring situations, such as monitoring engine bearings. Two types of RTDs are shown in Figure 4.18.

The thermistor is also small and is less expensive than the RTD or thermocouple. It is also used in moderate temperature ranges, but it is not as precise as the RTD. The thermistor is a semiconductor device whose conductivity changes with temperature. The change is not linear, so additional signal conditioning may be required. As you can see from Figure 4.19, they come in many styles.

▶ ***Optical Pyrometers Are Used for Noncontact Applications.*** Optical pyrometers use a focusing lens to direct photons onto a thermopile, which is a small bunch of thermocouples connected in parallel. The pyrometer is a noncontact gauge and is used in applications where contact sensors would cause physical damage. An example is to measure the temperature of a wet coating film on a moving substrate or web as the substrate exits an oven.

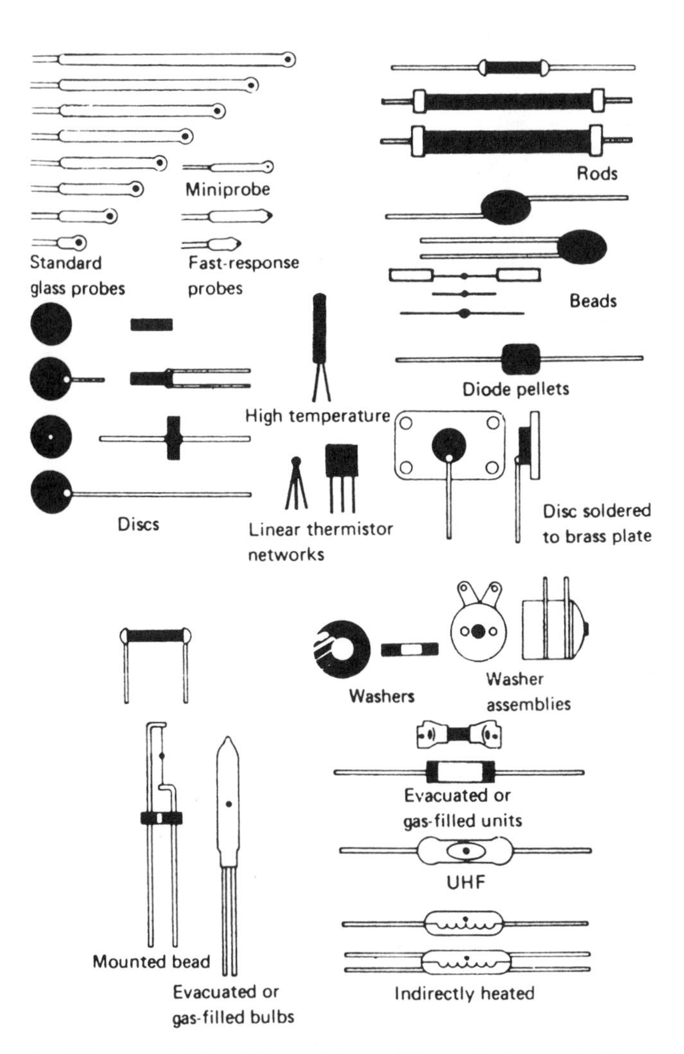

▶ ***Figure 4.19*** Thermistors (Humphries and Sheets, *Industrial Electronics*)

Nearly all temperature-measuring devices are available ready to use and most feature control capabilities. Set points with proportional control are common. Just about any kind of output is available, including digital in serial or parallel configurations.

Pressure and Flow Transducers

Pressure and flow transducers are used extensively in the process industries (e.g., in chemical plants), but less frequently in metal fabrication and assembly industries. Since the control function is usually more centralized in the process industry, we frequently find the transducer, associated electronics, and a terminal strip for hook up in the same package located at the measuring site. Such a device is called a pressure or flow transmitter, and early versions provided an analog signal of typically from 4 to 20 milliamperes (mA). Because of computer control and integration of the process industries, many more models are available with digital output.

Pressure is measured with strain gauge transducers and piezoelectric transducers, both of which we have examined, and by displacement measurement of a diaphragm. Usually, the displacement is measured with inductance or capacitance sensors so as not to contact the diaphragm. A pressure transducer is a type of load cell. In actuality, it is the force or load that is measured in a pressure transducer, and this force is divided by the area being acted on. Pure load cells are widely used to weigh things and to measure the forces exerted by hydraulic and other mechanisms.

Fluid flow is measured by measuring the displacement of a mechanism in the stream, by counting revolutions of a rotor or propeller, by the pressure drop across a constriction in the line, or acoustically. With acoustics we can sense the noise level made by the stream vortexes in turbulent flow by changes in sound transmission or with Doppler techniques.

Torque Measurements

Torque measurements, which are not usually done with a transducer, don't quite fit into a chapter on sensors. However, torque measurements are used extensively to determine tool presence and machine loading. In the process industries torque can be related to viscosity. Torque loading is usually determined by sensing the changes in amperage of the motor drive. If a cutter is missing, for example, motor current will be much less than if it were there. The method is not very precise because other things, such as bearing friction, can cause an increase in the current readings. Even so, some advanced systems use torque loading to provide feedback control signals.

Miscellaneous Gauges

Almost any physical phenomena can be measured and someone has a gauge for it. One characteristic of the instrumentation business is that physical principles are employed in many different combinations to achieve the desired objectives. The Instrument Society of America publishes a catalog, as do many of the instrumentation trade journal publishers. Some of the more common instruments available measure various kinds of radioactivity, moisture, pH, and the composition of matter. Some of the more exotic instruments may find their way into a CIM plant to measure such things as N_2 content in a treating furnace or SO_2 in a stack gas. A truly integrated CIM operation will be concerned about energy inputs and pollution monitoring as well as machine performance. From the point of view of the CIM engineer, it is likely that any gauge or instrument will have available a signal that can be digitized and an output that can be made linear through programming.

4.15 Data-Entry Terminals

▶ ***The Keyboard Is Widely Used but Is Error Prone.*** Data-entry terminals can be as simple as a telephone push-button pad or as complicated as a full function keyboard with local programming capabilities. Previously, all data were entered on a key punch card machine, but now most computers use a keyboard and monitor. The important point is that the most common people interface is still the typewriter keyboard. People are error prone, so methods must be incorporated to ensure proper data entry. One way to do this is to have a redundant system. That is, two people enter the same data. Each character is compared with its counterpart and the operators are alerted if parity (both characters the same) does not exist. Obviously, this is costly. Another way, but not as certain, is to have a checking system built into the program and documentation. This is called error trapping. If a part number contains five integers, the program should accept five and only five integers. Letters, punctuation marks, blanks, four numbers or less, and six numbers or more will cause a reject. This will not prevent a 3 for a 4 but will catch the letter O instead of a zero.

One reason there are big switches and buttons on industrial NEMA control panels is so that machine operators can use them with gloves on. Such working conditions might occur in a steel mill or coal mine. Conventional data-entry terminals that can be successfully operated on a foundry floor for any extended period of time are rare.

▶ ***Functional Illiteracy Is a Factory Fact of Life.*** There are many things wrong with keyboard entry systems, not the least of which is that the operator must be able to read. When well-educated people are entering data, errors are made, but at least the data get entered and can be corrected later. As data-entry terminals are used more in the unskilled areas of the factory, methods will have to be devised to facilitate the entry of data. Even if most unskilled factory help can read, many do not read English and most do not know how to type. For a nontypist, a keyboard is a very slow way to enter data. One solution has been to use function keys. Pushing "h" or "home" brings the tool back to the home or starting position. Many such ergometric challenges are being diligently worked on.

4.16 Tape Data Entry

▶ ***Tape Is Still King on the Factory Floor.*** The workhorse of data entry into computerized machine tools is still punched tape. Coordinate information and machine functions are programmed by the engineer into a computer with a FORTRAN type of language. The computer feeds a tape puncher that translates the digital information into code that comes out as a pattern of holes across a tape for each instruction. The tape is then delivered to the machine-tool operator, who puts it into the reader. The code is translated into machine operations like "move cutter head to X23, Y34 and turn on the coolant."

Floppy disk readers are now available, but their acceptance on the shop floor has been slow. In most factories the machine operators on expensive machines have seniority and are consequently older. These operators learned the CNC (Computer Numerical Control) systems with difficulty and are not prone to learning new methods, particularly since the tapes work so well. Some of the operators can look at the hole patterns and spot errors made by the programming engineer. Mylar and aluminum tapes are inexpensive, durable, nonmagnetic, and transportable. They can be folded, scrunched, and driven over by fork lift trucks and still made to work. Try placing a floppy disk on top of a large motor or other magnetic field. Older machines are being replaced with newer machines, which use the tape to load a small computer attached to the machine. In this case, the tapes can be made of less durable materials. Even newer machines are coupled directly to a host computer, which is usually programmed off line. This is called DNC (Direct Numeric Control).

▶ ***CIM Plants Will Be Directly Interconnected.*** The situation in a true CIM environment is different. Here we have eliminated the transition from computer to tape and back to computer and can now go directly from computer to computer. Methods are being put in place now in automated factories to ensure that the machines and computers are compatible and that machines operate according to a set of priorities and procedures. Transition to a full CIM operation will neither be quick nor painless, however.

4.17 Voice Data Entry

► ***Voice Entry Is the New Frontier.*** We can program machines to talk to us; just dial a nonworking number on the phone and see what happens. Not so simple is voice and language recognition by computers. Great strides have been made, particularly with physically impaired people, and soon we will see voice-recognition equipment appearing in the manufacturing environment. The problems are large, because every person can say the same thing but each will say it differently. Some people have high voices, some speak slowly, some slur their vowels, and so on. Technically, the problem can be solved, but it takes a great deal of memory storage and extensive program processing time. Most demonstration systems are slow. Even so, machines can now understand simple commands like "stop," "left," and "one, five, two." We are a long way from "Take the doohicky and put it in the thingamajig," and having the ¾-inch drill end up in the number one chuck. However, this is the objective.

4.18 Bar Coders and Other Semiautomatic Readers

If you have not yet seen a bar coder in action, you will soon. Many supermarkets and other retail establishments have them at the checkout counter. Imprinted on the package or box in a special location is a series of black and white lines. At the checkout, a laser beam scans the lines and the light is reflected differentially. Optical sensors in the machine can pick up the pattern, decode it, and output the information digitally. Figure 4.20 shows how the laser coder operates. Industry has been slow to take advantage of this technology, but things are changing quickly. Now there are many makers of bar coding equipment and several types of codes. It is hoped that standards will soon be established.

► ***Bar Coding Use Is Accelerating as a Management Aid.*** Bar coding is now used for inventory and production control. In a typical setup, a part will have a sticky-label bar code placed on it directly or placed on the traveler envelope that goes with it. At each location at which the part stops, the code is read and entered into the computer system. In addition to identifying the part number, bar codes can be used as a serial number for the individual part. In this way the control department knows at all times where the part is. If there is a delay in the process, the computer can alert the manager so that the cause can be quickly determined. Quality assurance information can be entered into the computer at each station to provide an accurate record of production quality. The system is rapid and certain, assuming, of course, that data are entered correctly. Although, initially, bar-code readers

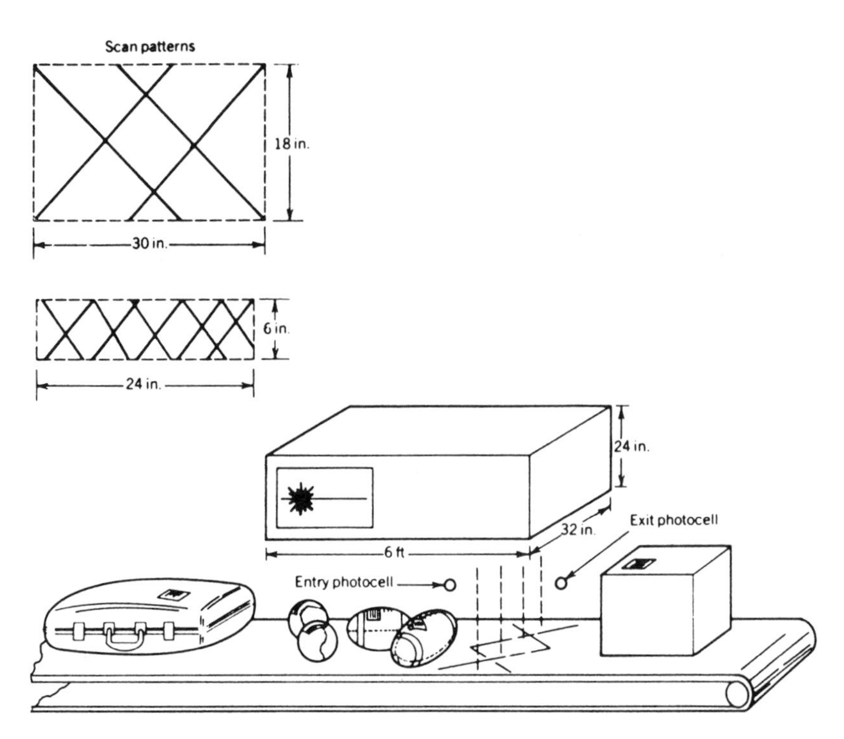

The laser scanner has large depth of field and uses a crosshatched scan to identify bar code labels of random orientation and distance from the scanner.

▶ ***Figure 4.20*** Laser Scanner (Asfahl, *Robots and Manufacturing Automation* and Rexnord Inc.)

were expensive, they now cost about the same as any other data-entry terminal. As the data base and computer capability grow, so will the use of bar coders.

▶ ***There Are Many Codes.*** The use of bar coding to facilitate inventory and production control is growing rapidly. Like everything else in the computer industry, there is no standard bar-code system, although four codes are used more frequently than others in a factory environment. These are Interleaved Two of Five, Code Three of Nine, Code 128, and the ubiquitous UPC, which is more familiar in supermarkets. Some of these can only read numerals and others can read both numerals and letters of the alphabet. Figure 4.21 shows what some bar-code formats look like. The format selection depends mainly on the intended use of the system.

The size or spacing of the bars on the code is important, with bigger being better. Here the problem is resolution. The quality of printing is important in order to provide sufficient contrast. The code

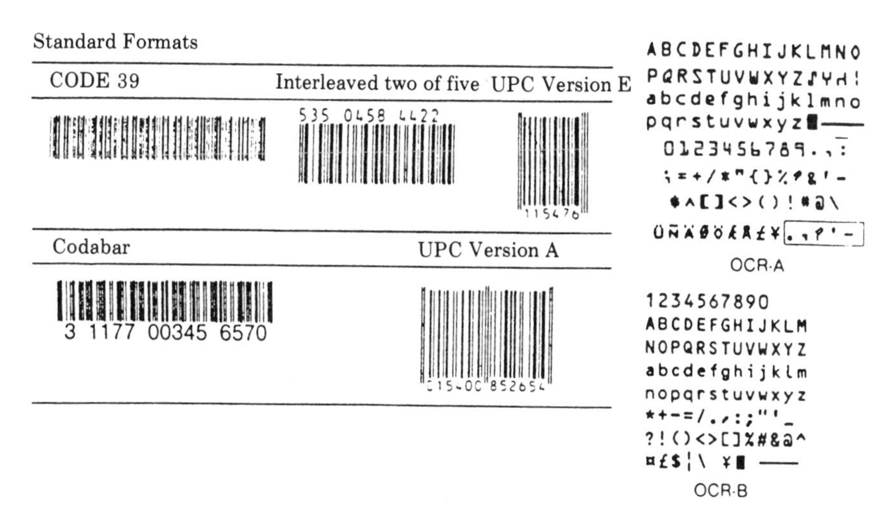

Shown are examples of standard bar code and optical character recognition formats.

▶ ***Figure 4.21*** Typical Bar Codes (Asfahl, *Robots and Manufacturing Automation* and Metrologic Instruments)

reader can be frustrated by the bar code being abraded off due to traveling throughout the plant or definition can be lost because of poor ribbons on the printing machines. The nature of the substrate upon which the code is fixed plays a part also. Assuming the code is affixed properly, code reading is very fast and the reliability is high. A key disadvantage of bar coding, however, is the inability of people to be able to read the codes. This slows up the process when you want to know what is in a box and have to take it off line and bring it to a reader. Like everything else, it takes careful planning to ensure that the coding system is cost justified and will be operational as intended when implemented.

Other Systems

Other identification systems have been developed, principally magnetic tape and alphanumeric cipher optical readers. Whereas bar coders can be read in almost any attitude or orientation, magnetic and character readers must be positioned fairly accurately relative to the sensors. Most systems have an alarm to alert the operator when the data do not register. Sometimes the line or process halts until the data are entered properly, and in other cases the part will be physically rejected.

4.19 Summary

We need I/O devices in order for the computer to be able to receive information from people, sensors, or other computers and to provide control signals and data to equipment and other computers on the outside world. From the computer's viewpoint, every I/O has its own memory address. I/O devices or peripheral equipment are connected to the computer through interface systems.

Input devices include such things as terminals and keyboards, card readers, bar scan coders, and a wide variety of sensors. These include contact and noncontact position sensors, encoders, vision sensors, and sometimes process instruments such as temperature, force, pressure, and flow sensors. Since most sensors are analog, an A/D converter is usually required, along with signal-conditioning electronics. Output devices include stepping motors, servo motors, and ac and dc drive systems for hydraulics and pneumatics.

The I/O function can be improved by making the system less dependent on people for data entry. Bar coders are starting to be used extensively, as are optical character readers. Voice data entry is being extensively researched, but there are many problems. Many applications are beginning to use computers at a higher level in the hierarchy. When this happens the factory will be more completely integrated.

4.20 Exercises

1. What do we need in a manufacturing environment in addition to what we find in a business office in terms of computer equipment?
2. Can two pieces of apparatus have the same address?
3. What two types of controls can be used?
4. What is the trade-off in robot arm movement?
5. Where is hydraulics used in driving robots?
6. What is a major problem regarding operating computer devices in a factory environment?
7. Who is responsible for checking the machine design?
8. What is the most widely used type of motor in the factory and why?
9. What is the advantage of dc motors?
10. What is the difference between phase angle fire and crossover fire?
11. How does a stepper motor work?
12. What are two types of encoders?
13. What does a servo device do?
14. Describe the action of a servo motor.
15. What do A/D and D/A converters do?

16. How do A/D and D/A converters work?

17. Why are visual and tactile sensors used in robots?

18. Basically, what two types of visual sensors are employed?

19. Describe the silhouette method of determining a part.

20. What two types of tactile sensors are generally used and what is the principle of operation of each?

21. How does a piezoelectric device work?

22. What is the principle of operation of the sonic sensor and where does it come from?

23. What is the most popular type of transducer and how does it operate?

24. Name several kinds of noncontact gauges.

25. What is meant by resolution?

26. Name several types of temperature sensors.

27. Name some other kinds of transducers that are used in the process industries.

28. What does a load cell do and how does it work?

29. How do we usually enter data into a computer?

30. Describe a redundant system and what is meant by a checking system?

31. What is a major problem in getting data into the computer from the factory floor?

32. Why are punched tapes still used extensively on the factory floor?

33. What are some problems with voice data entry?

34. Describe the operation of a bar coder.

35. Name three different bar code systems and explain why there are different ones.

36. What is the disadvantage of the bar code with respect to interfacing with people?

5

CIM Units: The Robot

5.1 Introduction

This chapter deals with a specific application of robotics, that is, the robot. Robots use the robotic hardware and software discussed in earlier chapters. However, the robotic hardware and software can be used in many applications other than robots. A programmable automatic guide vehicle (AGV) is an example of a machine that uses robotics but is not a robot. Robots are of particular interest because they employ, in one handy package, most of the key components of robotics. An understanding of how a robot works and what it is good for will pave the way for understanding other robotic devices. It is conceivable to have a totally integrated CIM facility, with all sorts of fancy robotics, and not have a robot anywhere on the premises.

► ***Robots Are Well Known.*** Much has already been said about robots. Robots have been covered extensively in the popular press and in-depth courses are offered in our universities. Robots have received wide coverage because they are so easy to talk about. If we watch a computer at work, we don't see much, perhaps a few blinking lights and reeling tape. A robot, on the other hand, has motion with rhythm and, in many cases, grace. Even if we don't understand them, they are fun to look at. Photo 5.1 shows a typical robot for medium-duty applications. Robots are intended to operate in places where humans exist, and so it is easy to love or hate them depending on your point of view. This is why robots are almost always given a personal name by the people in the factory. We identify with them personally because they have some human characteristics.

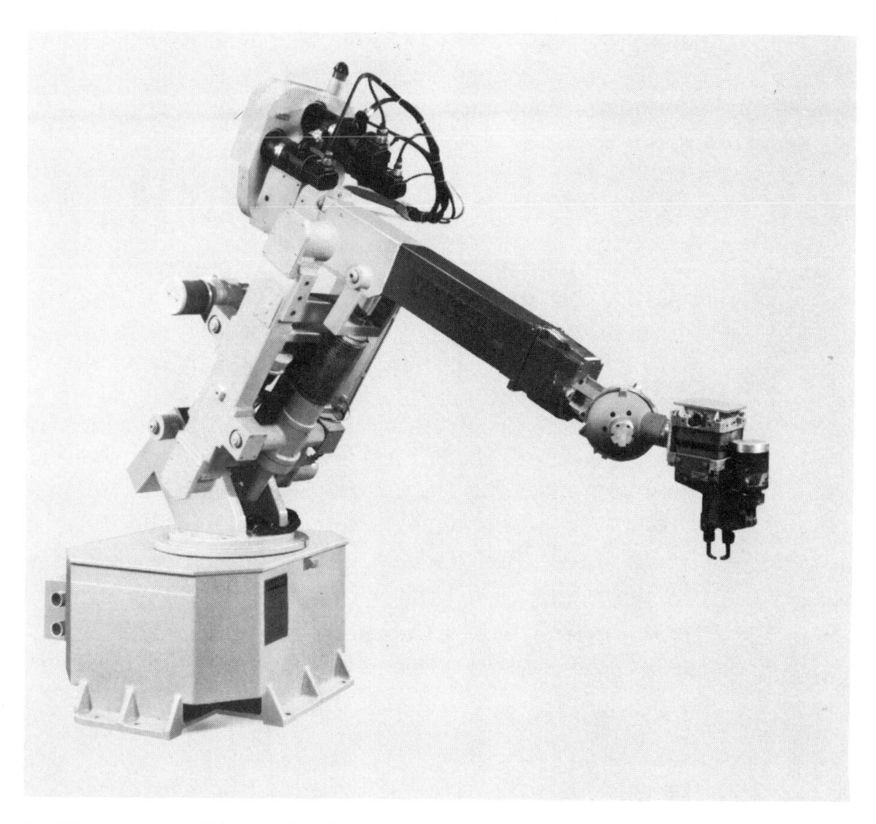

▶ ***Photo 5.1*** Heavy-Payload Robot (Courtesy of Industrial Robot Division, Cincinnati Milacron)

5.2 *Definition*

Robots Defined

The Robotics Institute of America defines a robot as a "Reprogrammable multifunctional manipulator designed to move material, parts, tools, or specialized devices through variable programmed motions for the performance of a variety of tasks." This definition seems to be adequate and is the one used in this text. Bear in mind, however, that other equally credible groups and organizations may have different definitions. Most notable is the Japanese Robotics Institute.

Manipulators and Degrees of Freedom

"Reprogrammable" and "variable programmed motions" imply some sort of programmable microprocessor or computer. "Multifunctional manipulator'' means that the manipulator must be capable of more than 1 degree of motion. Many people feel 3 degrees of motion is the minimum for manipulator capability in order to qualify in robotic terminology. This would exclude a simple pick and place device, which has only 2 degrees of motion. The Japanese, who are more liberal in this regard than we are, do not use the Robotics Institute definition and count things as robots that we do not.

Something has to be moved from one place to another. ''Material, parts, tools, or specialized devices'' covers just about everything one can think of.

The RIA definition sounds simple enough. Is an electromechanical tool that is reprogrammable and that can operate in any attitude a robot? One specific case is an electrical discharge machine at General Electric that meets all the criteria of the RIA definition, yet neither the machine-tool builder nor the owner count it as a robot. On the other hand, a reprogrammable component insertion machine is sold as a robot even though it has limited degrees of functionality. Figure 5.1 shows a cylindrical coordinate robot and its principal parts.

► ***RIA Definition Is Not Complete Nor Accurate.*** For the above reasons, we find wide disagreement on how many robots are in use and

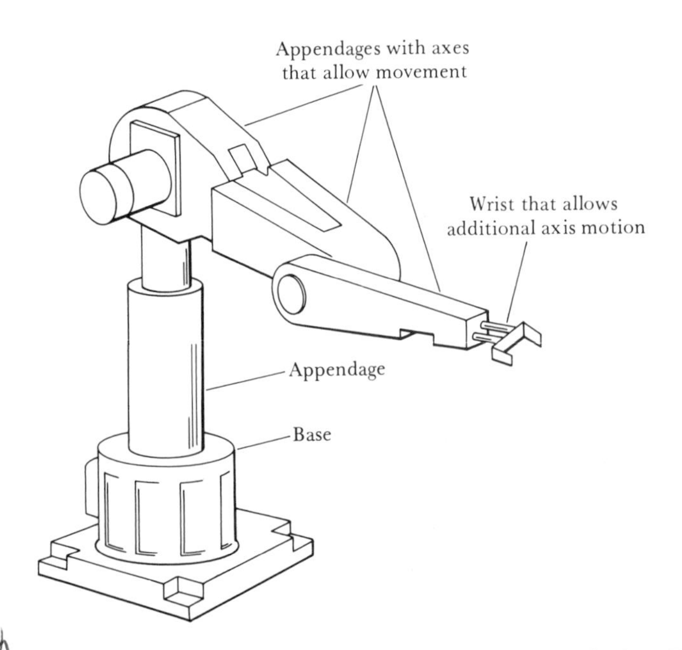

► ***Figure 5.1*** Principal Parts of a Robot (Malcolm, *Robotics*)

what the market is. If you think a robot is good, you will call your machine a robot. If you think a robot has a negative connotation, you will call it something else. Programmable controllers were called programmable controllers because it was feared the term "computer" would create union problems in Detroit, where they were first introduced.

▶ ***Number of Robots in Use.*** By consensus, there were about 20,000 robots in the United States in 1985, which seems to be a reasonable figure. Worldwide figures given in Figure 5.2 show there were 28,402 robots in 1983. If you look at projections for 1985, which were made in 1980, you will discover that robots have failed to live up to their market potential by a significant amount. Forecasters were too optimistic. Harder to guess is what the market will be in 1990 and 1995. Projections for 1990 range from about 75,000 to 100,000 units, quite a wide discrepancy. About the best that can be said is that many more robots will be used in the future. Less costly machines, wider applications, and better servo controls will be growth factors. Interesting from a different perspective is the fact there are more toy robots than real ones.

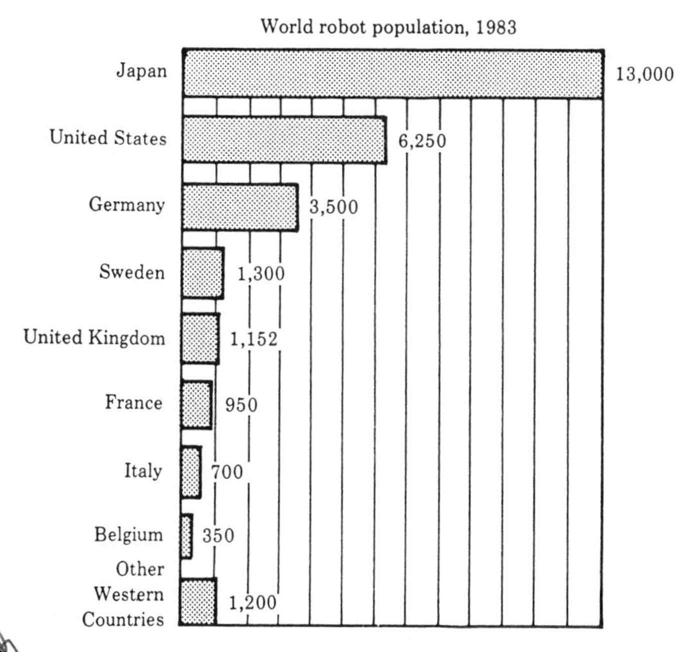

▶ ***Figure 5.2*** Robots in Use

5.3 Types and Classifications

Robots can be classified by function, by manipulator arm geometry, and by complexity. These are not the only classifications, but they will do for now.

Robots Classed by Function

Currently, robots are being used mainly in locations where it is unattractive or hazardous for humans to be. In theory, any job can be made attractive or can meet minimum safety standards if we spend enough money. The question then becomes an economic one. Robots are being used for material handling, welding, painting, loading, assembly, and other applications such as adhesive placement, cutting, and deburring. Since a robot, once programmed, travels over the same prescribed path and performs its operations in programmed sequences, the finished product varies much less than products made by humans. A good example is welding. We see a dramatic presentation of a welding application in Photo 5.2. Who could make consistently good welds hour after

▶ ***Photo 5.2*** Robot Welding (Courtesy of Industrial Robot Division, Cincinnati Milacron)

hour in all kinds of odd positions wearing an uncomfortable face mask and enveloped in noxious fumes?

The most widely used robot is the simple pick and place robot. Machine loading, particularly where a hazard exists such as in die casting or forging, accounts for the majority of these in the metal industry. Component insertion in electronic circuit boards is also a very large user of pick and place robots. Component insertion is usually thought of as an assembly operation. Robots are not too widely used in assembling things where manual dexterity and a high degree of eye–hand coordination are required. Only a few robot manufacturers specialize in general-purpose robots. Most makers are associated with a particular process or industry. Hobart Corporation specializes in welding robots mainly for the automotive industry, whereas Teledyne-TAC specializes in assembly robots for the semiconductor industry.

Robots Classed by Arm Geometry

Robots can be classified by the manipulator arm geometry. All manipulators use either polar or Cartesian (XYZ) geometry or a combination thereof. Some people break down these types further into Cartesian, cylindrical, polar, and articulated types. Figure 5.3 shows the more popular configurations. For example, a robot may have a cylindrical base to move about a central axis for left to right motion (polar) and employ in/out and up/down motion (Cartesian) for the arm. The proper choice will depend on the application.

▶ ***Envelope of Operation.*** An important consideration is the operational envelope in which the end of the manipulator travels. In general, the articulated-arm robot (most people's idea of a robot) can cover more territory with less structure than other types for the same load. This is why we see so many of this type. Figure 5.4 shows what we mean by envelope of operation.

For uniform precision within the envelope, the Cartesian system has a distinct advantage. A Cartesian system configured as a gantry robot may be able to position to, say, one-thousandths of an inch (0.001) anywhere in a 10-foot cube. This is not too difficult to achieve. A polar system, on the other hand, must be able to resolve the angle at the axis to increasingly higher orders of magnitude as the size of the envelope increases in order to reach the same precision. A 0.001-inch precision at the end of a 10-foot arm requires an encoder that can resolve to 1 part in 120,000. This is very difficult and expensive.

▶ ***Geometric Considerations.*** Trigonometric and geometric relationships in Cartesian systems are certainly simpler than in polar systems. The drawing in Figure 5.5 gives some idea as to the kind of motion. Try on your own to develop the mathematic formula for the straight-line path of the end of an articulated arm and you will get an

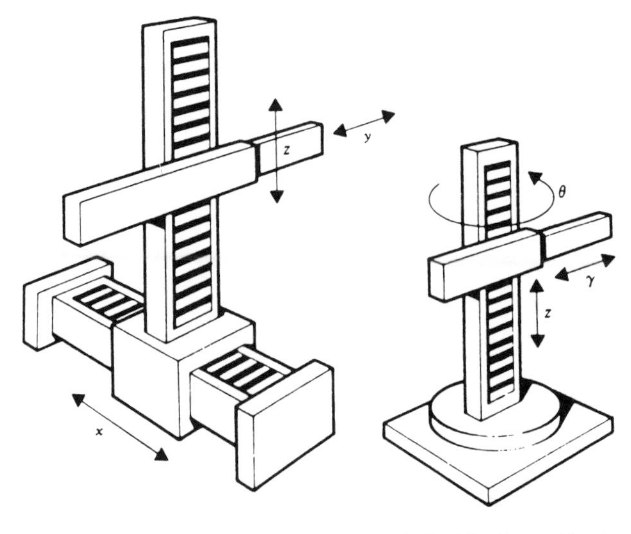

A. Cartesian coordinates B. Cylindrical coordinates

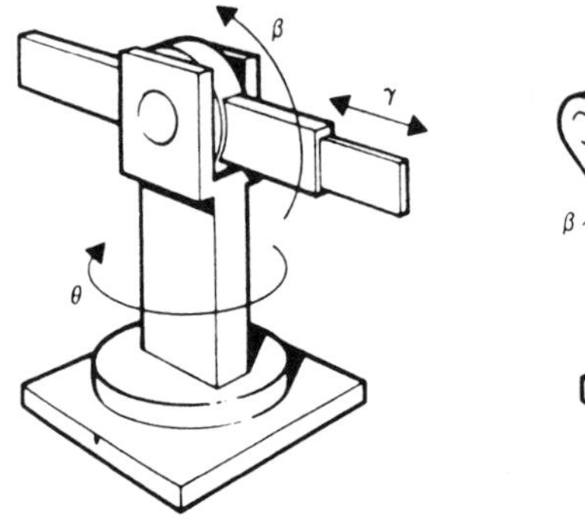

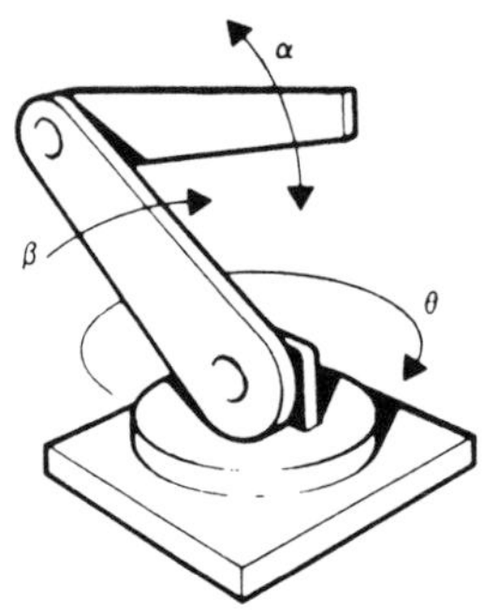

C. Polar coordinates D. Revolute coordinates

▶ ***Figure 5.3*** Robot Arm Geometry (Engleberger, *Robotics in Practice*)

idea of what is involved. Fortunately, most of these relationships can be incorporated into the microprocessor design as firmware or through extensive programming subroutines.

▶ ***Meaning of Degrees of Freedom.*** Robots can be classified by their degrees of motion or freedom. Degrees of freedom refers to the motions the machine can undergo. In/out, up/down, and right/left are the three basic motions and are considered the minimum required to meet the criteria of multifunctionality by RIA standards. To these we can add wrist motions, such as yaw, pitch, and roll. If there is a gripper,

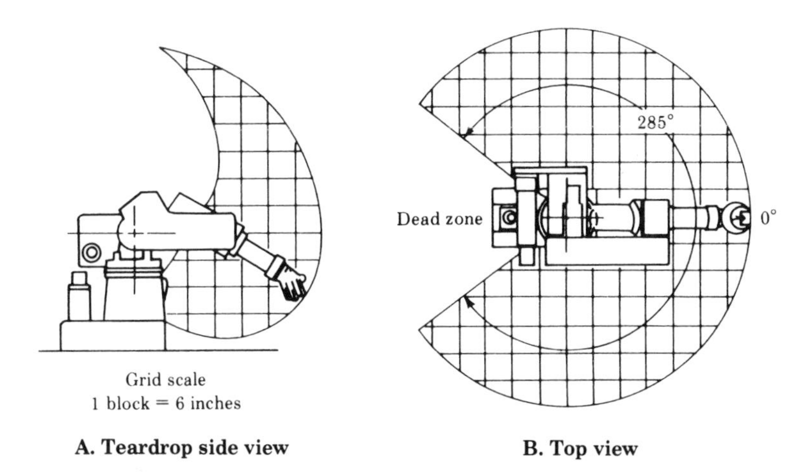

▶ ***Figure 5.4*** Typical Robot Arm Envelope (Malcolm, *Robotics*)

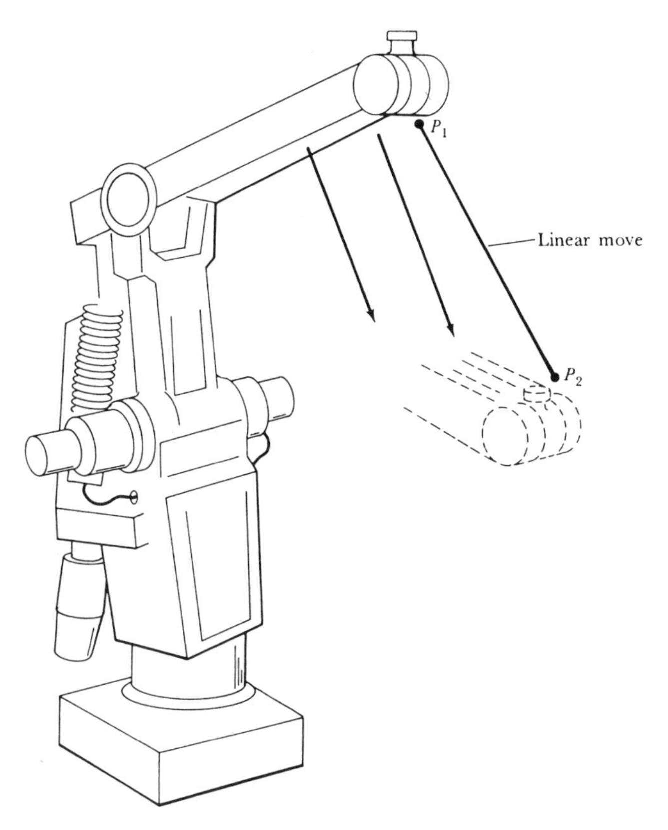

▶ ***Figure 5.5*** Robot Arm Path (Malcolm, *Robotics*)

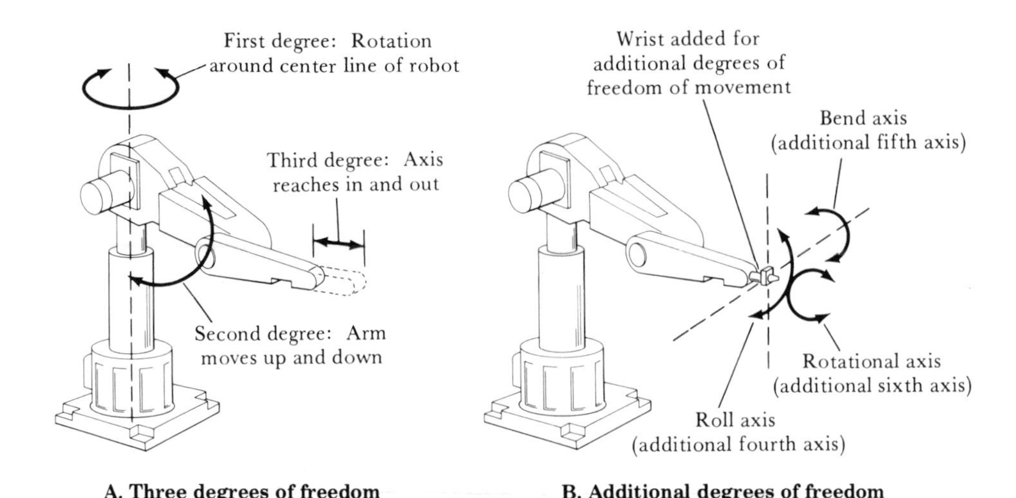

▶ ***Figure 5.6*** Degrees of Freedom (Malcolm, *Robotics*)

it must open and close. Figure 5.6 shows what is meant by degrees of freedom. All told, more than 15 degrees of freedom are possible, but most robots employ only 6 to 9.

Robots Classed by Complexity

We can also look at robots from the point of view of the technology employed. Some robots are very simple and do not have much in the way of technology. A simple loader/unloader for a forge may have only 3 degrees of freedom, be Cartesian, be precise to 0.01 inch, carry light parts, and employ an open-loop control system. Such a machine would be relatively inexpensive. A large-payload, articulated-arm, polar-coordinate robot with high precision, many degrees of freedom, and fast response, such as might be found on a paint spray or welding line, employs a great deal of technology and is expensive. Depending on the complexity, we generally speak of low-, moderate-, and high-technology robots, but the boundaries between these classes are indistinct. A low-technology robot system is shown in Figure 5.7. In general, the more complex and precise the robot is the more it costs. At present (1987), $15,000 will buy a pick and place robot with modest capability that is able to handle a 5-pound load, whereas a paint robot costs a minimum of $60,000.

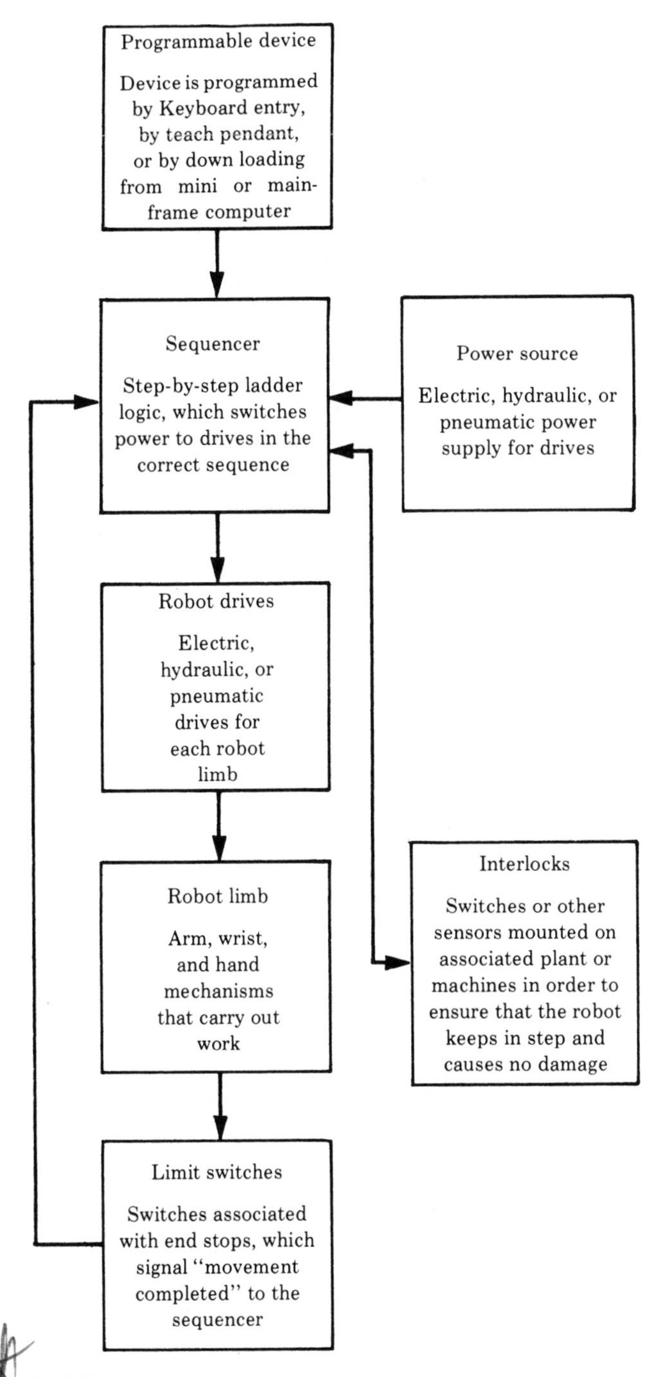

▶ ***Figure 5.7*** Low-Technology Robot System (Engleberger, *Robotics in Practice*)

5.4 Components

Power Supply

A robot is a machine tool. As such, it needs a power supply for the electronics, as well as for the power mechanisms. Depending on the size and complexity of the robot, the power mechanisms' power supply can be a simple transformer and rectifier to drive the small dc stepping motors or a very large ac-driven hydraulic unit with complete starter and overload circuitry. The power supply for the controller portion is always isolated from the rest of the electrical equipment to protect it from noise. The integrated circuits employed are very sensitive to transients and static electricity and are easily damaged. Even if "glitches" do not cause damage, they can easily change what is in memory, and the resultant loss of data or programming creates sporadic operation. Most TTL-compatible circuitry uses +5 volts dc, and much of the I/O ports are either +12 or +24 volts dc. Enough current-carrying capacity should be available to handle not only the requirements of the controller but also any add-on equipment that might be connected to the I/O ports or motherboard.

Electronics in a Robot

In addition to the power supply, the controller contains all of the other components found in a typical programmable controller or microprocessor. Diagrams of the internal components of a robot are shown in Figure 5.8. The motherboard is a printed circuit board that usually holds the basic components of the processor. Here we find the microprocessor unit (MPU), which does the computation and sends the information to the various address locations. Also, the system ROM, which contains the built-in programming and machine subroutines, and enough RAM, to store the data generated by the programs are on the motherboard. Sometimes we find additional RAM, I/O circuitry, servo control circuitry, and other interfacing circuitry on the motherboard. However, the usual practice is to place PCB connectors on the motherboard so that the designer of the system can add whatever components are required as separate PCBs, depending on the configuration of the system. The motherboard also contains buses or lines so that the various components can communicate with one another. A bus arrangement is given in Figure 5.9. Most robot makers are simply assemblers of other people's components. The controller is, therefore, a purchased part from a maker of programmable controllers or microprocessors.

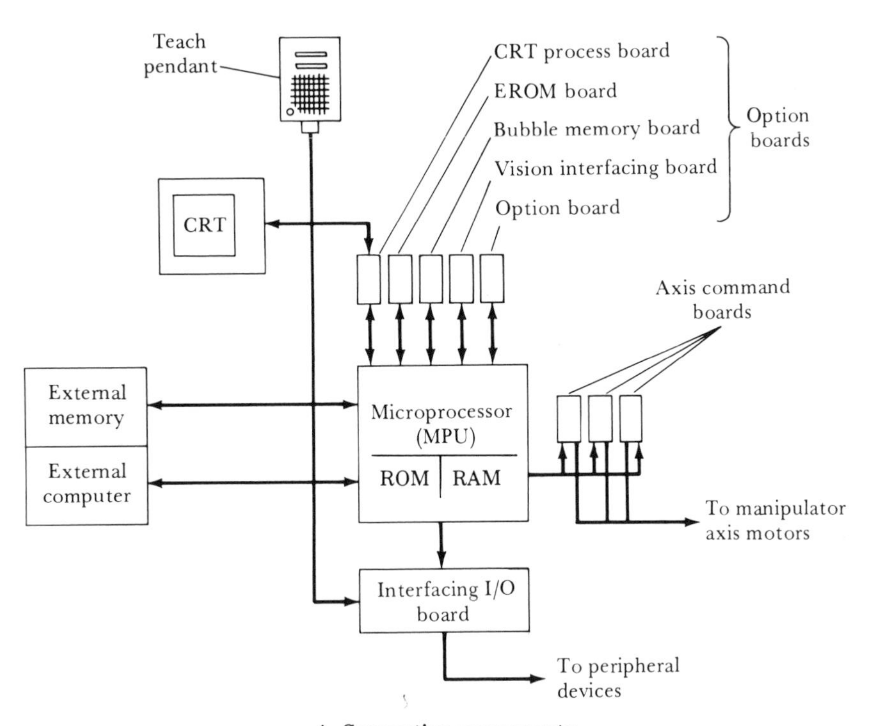

A. Supportive components

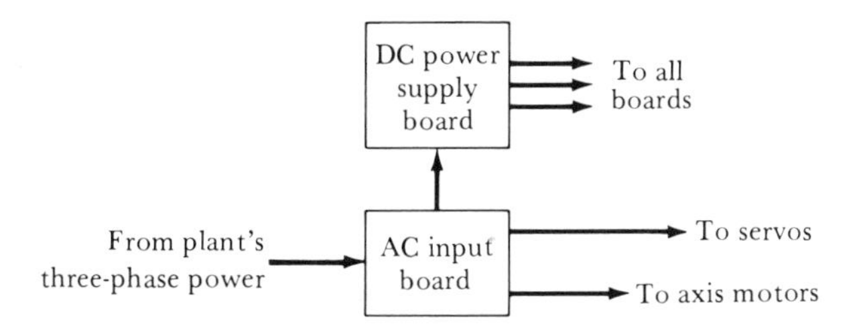

B. Power supply board

▶ ***Figure 5.8*** Internal Components of a Robot (Malcolm, *Robotics*)

Control Panel

Mounted on or near the robot is the control panel. Figure 5.10 shows a typical control panel. In addition to the on/off switch and power lights, we usually find controls to put the robot into several modes. When in automatic mode, the controller is driving the robot. We also need to be able to position the robot manually for setup and to move it out of the way when required. When in the manual mode, the operator can control the robot

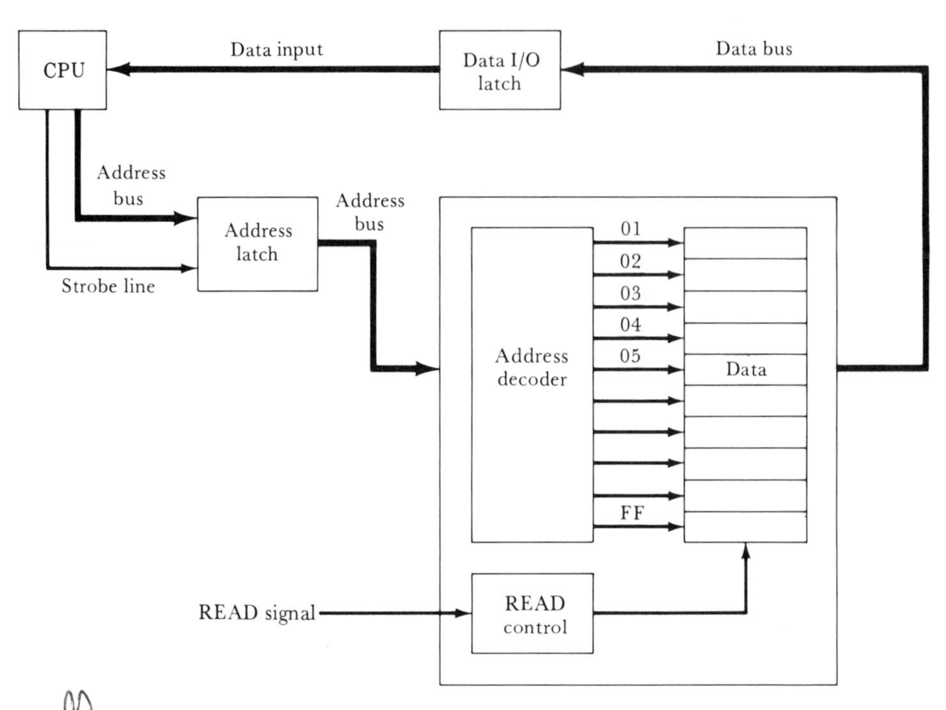

► ***Figure 5.9*** Typical Data Bus Configuration (Malcolm, *Robotics*)

through the use of buttons and switches mounted either on the panel or contained in a separate box connected to the machine by cable. More often than not the separate box is also used as a teach pendant.

A robot teach pendant is shown in Figure 5.11. It enables the operator to "jog" the robot into proper coordinate position through a series of small steps. Once at the desired location, the operator can set the coordinates and velocity information into the controller as a programming step. This procedure saves much time because the operator does not need to worry about the intricacies of programming, languages, syntax, and so forth, but can concentrate on the job of getting the robot to go through the proper motions. The program, once learned, is stored in memory or on disks for future use. Photo 5.3 shows all the components of a Unimate robot system.

Slippage and Backlash

Whether the robot is small or large, simple or complex, or polar or Cartesian determine how the manipulator or end effector arrives at its proper destination. Smaller robots employ electric motors and actuators. Large robots employ hydraulics and pneumatics. All use mechanical linkages and power trains to transmit power to the appropriate

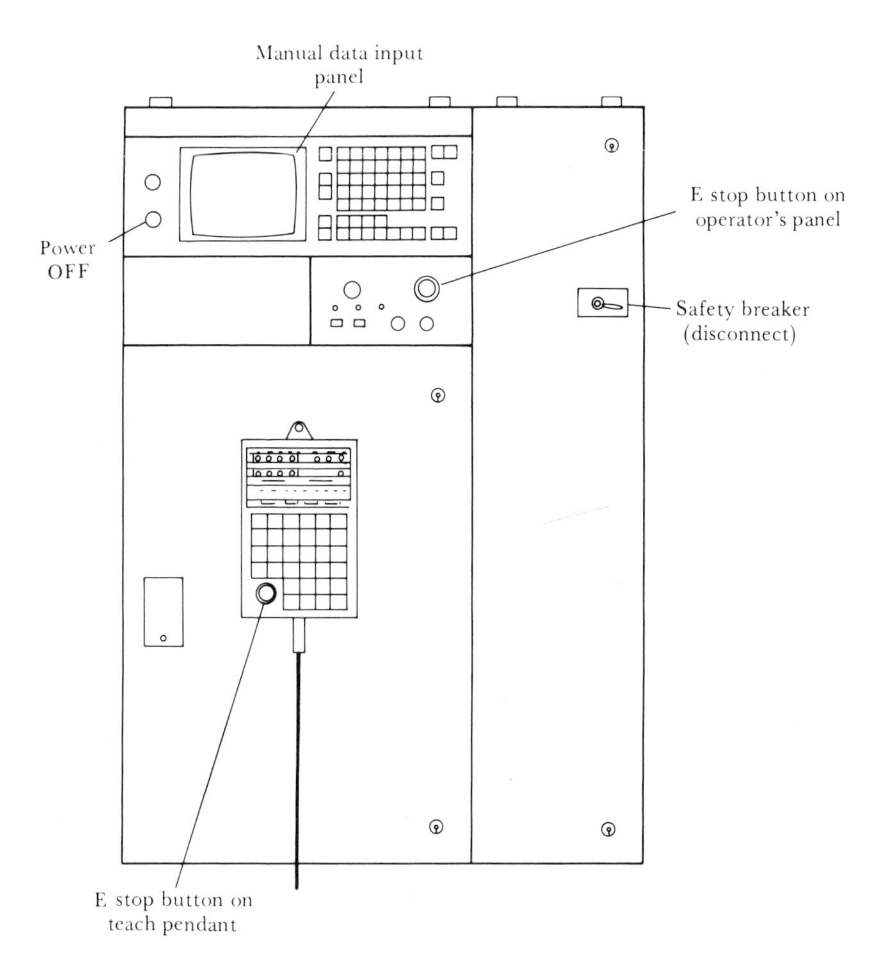

A. Location of operator's panel on controller

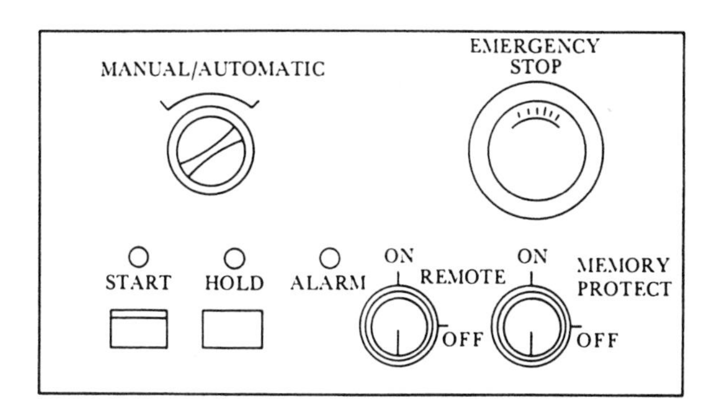

B. Detail of operator's panel

▶ ***Figure 5.10*** Typical Robot Control Panel (Malcolm, *Robotics*)

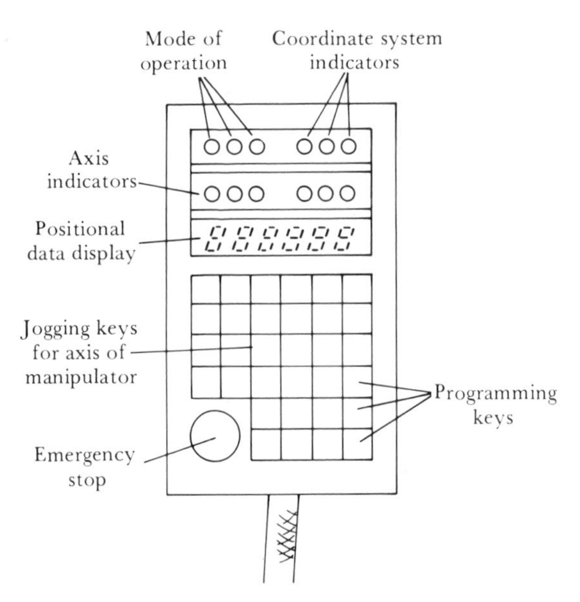

A. Components of the teach pendant

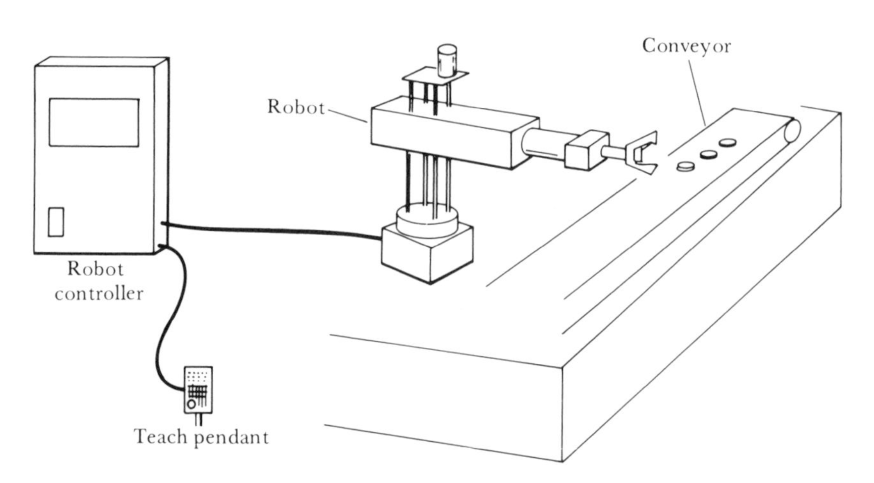

B. Teach pendant brought into the work envelope

▶ ***Figure 5.11*** Robot Teach Pendant

places. Great care is taken to ensure that the arm or tool holder reaches the proper place at the proper time. However, all power mechanisms have either slippage, backlash, or both, and the degree of precision depends on how little slippage and backlash are found in the components of the power train. Slippage occurs when friction or other forces inhibit

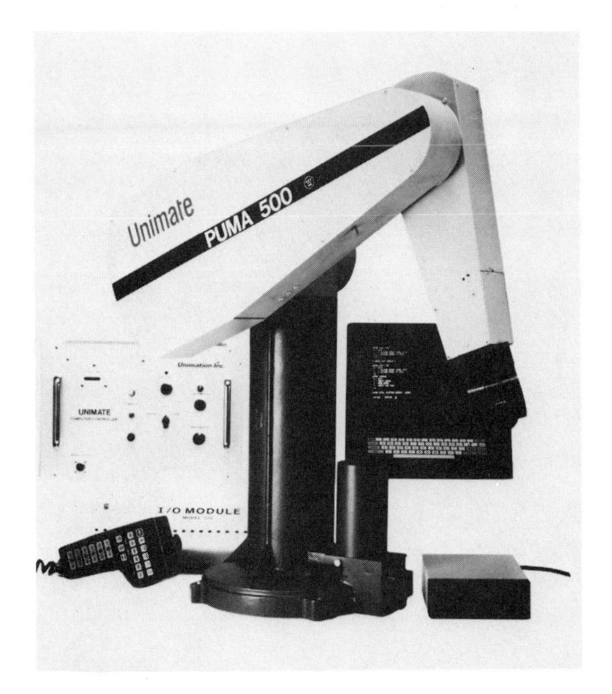

► ***Photo 5.3*** Small Parts Handling Robot (Copyright UNIMATION Incorporated, A Westinghouse Company)

motion. For example, a hydraulic motor uses a noncompressible (for practical purposes) fluid to drive it. If the machining is perfect, no fluid will leak around the pistons, gears, or vanes. However, such perfection is an unachievable goal. A pneumatic cylinder that has a great deal of internal friction due to improper lubrication or dirty air will not reach the end of its stroke. Dirt coupled with motion causes erosion and wear and is the enemy of anything mechanical, because erosion and wear cause friction and slippage.

Backlash occurs when there is imperfect contact between two components in the power train. The diagram in Figure 5.12 shows the idea of backlash as it relates to gears. We can make the gears with very tight tolerances to greatly reduce backlash, but then wear will be excessive.

Use of High-Tech Components

Excellence in the maintenance program will do much to solve wear problems. The time for proper maintenance is before failure. Backlash can be reduced by proper design and selection of components. Vendors who supply mechanical and electromechanical components strive to provide products with very high degrees of built-in technology. It is unfortunate that the term high-tech is understood by most people to

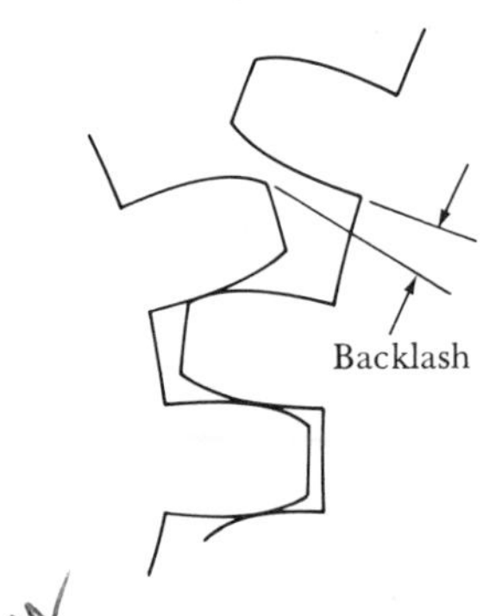

▶ ***Figure 5.12*** Gear System Backlash (Malcolm, *Robotics*)

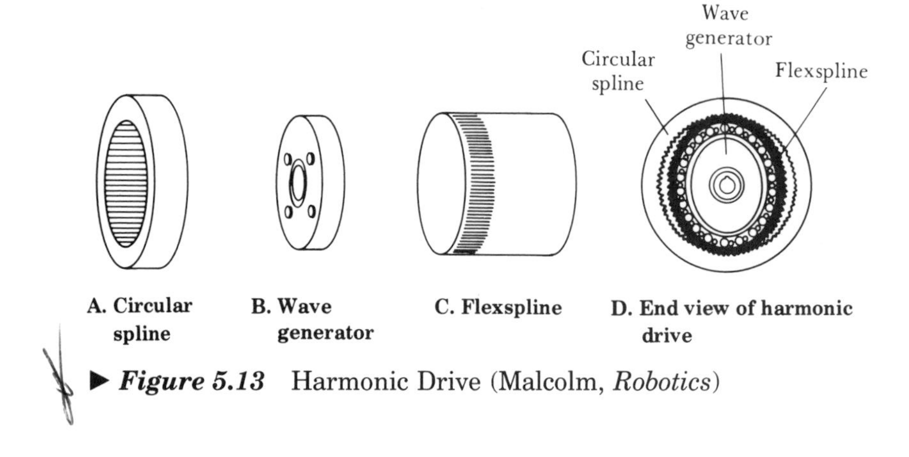

▶ ***Figure 5.13*** Harmonic Drive (Malcolm, *Robotics*)

refer to computers, for there is plenty of high-tech in the mechanical field. A mechanical example is the harmonic drive. In essence, this design uses a flexible gear that is always forced to mesh regardless of motion, and thus eliminates backlash. Other examples are the ball screw drive, which uses traveling ball bearings as part of the threads, and the five-pole stepper motor, which gives better resolution than two-pole motors and has almost eliminated problems due to resonance. Figure 5.13 shows the principle of operation of the harmonic drive, and Figure 5.14 shows the ball screw mechanism.

Feedback and Servo Control

One way to ensure that the robot reaches the correct position is to use adjustable mechanical stops. This method is widely used on pick and place robots, but it has the disadvantage that the stops need to be moved for each setup. Another way is to use servo mechanisms. A schematic of

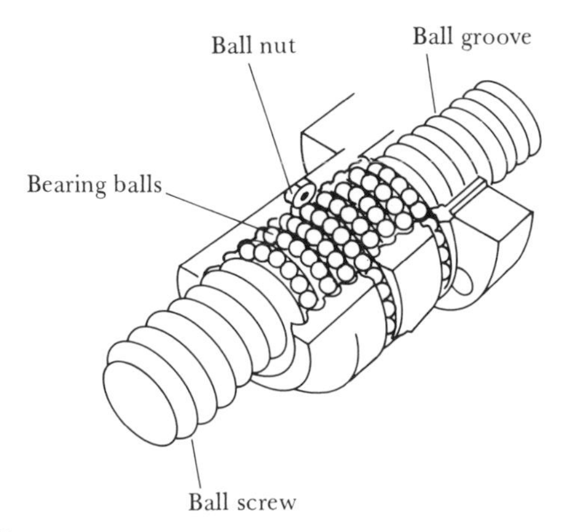

▶ ***Figure 5.14*** Ball Screw Mechanism (Malcolm, *Robotics*)

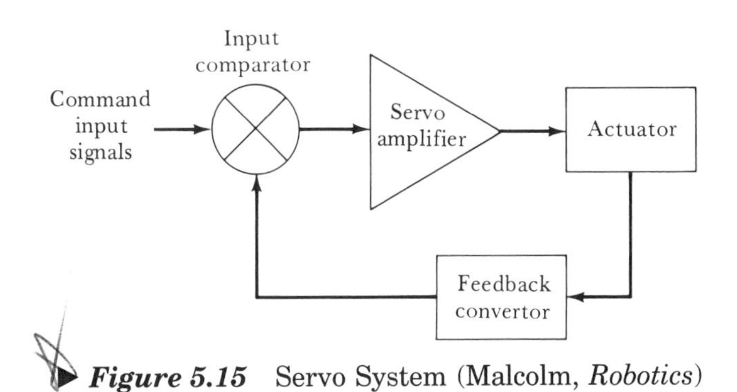

▶ ***Figure 5.15*** Servo System (Malcolm, *Robotics*)

the servo concept is given in Figure 5.15. The use of a feedback loop is the key ingredient of a servo system. The position of the arm is measured by independent means. The processor tells the arm where to go, and the arm will approach the proper coordinates. The difference between where the arm actually goes and where the processor sends it is measured. This difference is called an error signal. The error signal is fed back into the controller through a feedback loop. The processor provides an incremental correction to the arm proportional to the error signal, which moves the arm closer to the target. A new error signal is generated, which causes the arm to move even closer, and so on. Sensors, such as were described in Chapter 4, are required. They provide the necessary error signal. Proper servo control is straightforward in concept, but hard to achieve in reality. In Figure 5.16 we see a diagram of a typical complete servo control circuit.

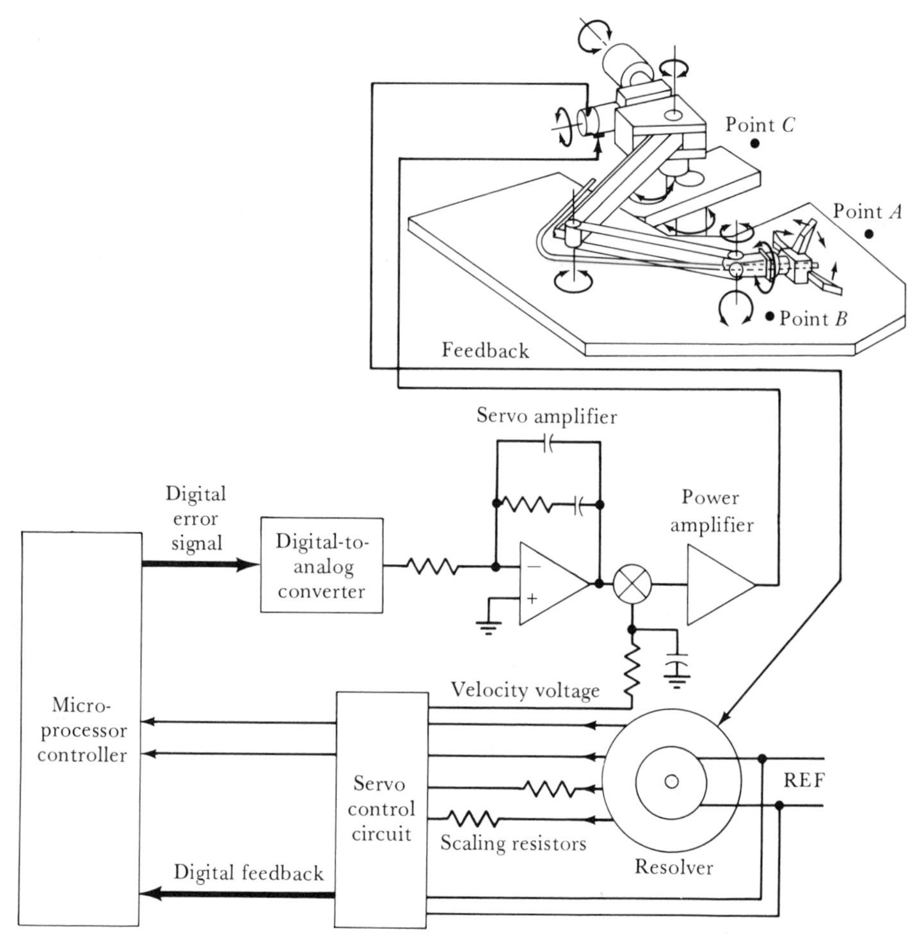

▶ ***Figure 5.16*** Typical Control Circuit (Malcolm, *Robotics*)

Control Characteristics

Ideally, during each loop the arm travels toward the target, starting with large corrections and zeroing in on the target until the target is hit. This should proceed smoothly and rapidly. However, things can go wrong. The arm may never reach the target. This is not likely to happen with a stationary target, but it happens frequently in conveyor situations. Sometimes the arm moves to the target and overshoots it somewhat due to an overcorrection signal. The arm then travels too far in the other direction, resulting in a dithering situation in which the arm oscillates back and forth. If conditions are such that on each successive oscillation the arm gets closer to the target, eventually the target will be hit. Some control systems are designed this way intentionally and are said to be damped. The advantage of this type of control is that you are certain to reach a resting point that is within the range of acceptable upper and lower limits. Many robot applications are such that a small amount of overshoot can be tolerated. If the arm continues to oscillate back and forth without ever reaching the target, the system is said to be hunting. In a catastrophic situation, the arm moves in ever-increasing oscillations until self-destruction or other calamities occur. This does not usually happen, however, unless there is a malfunction or component failure within the control system itself.

Precision and Accuracy

Precision or repeatability is the ability to hit the same spot on the target every time. Closed-loop servo systems that employ encoders for positional determination on the robot itself do a good job of maintaining precision. Of course, the servo system must be properly designed and adjusted in the first place so as to avoid the conditions already described. However, precision and accuracy are not the same thing. Accuracy is the ability to hit the bull's-eye of the target. The tighter the grouping of hits in the bull's-eye, the greater the precision is. But the tight grouping may be up in the left-hand corner of the target and nowhere near the bull's-eye. This is good precision but poor accuracy. To ensure accuracy, we need to measure positional information with systems that are not coupled directly to the robot drive, but are separate unto themselves and can be referenced independently. Figure 5.17 shows the difference between precision and accuracy using a rifle target.

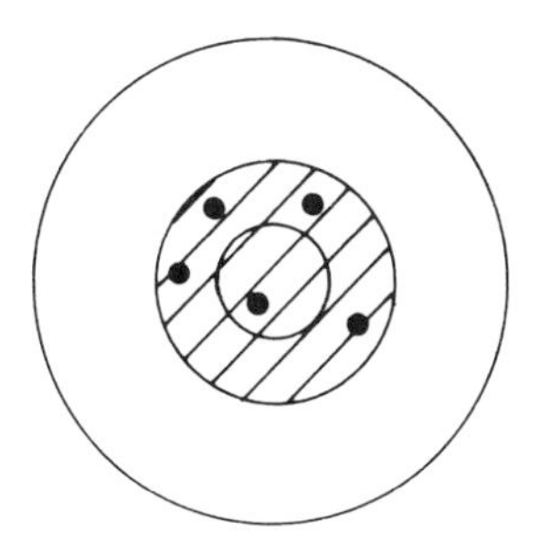

A. Accurate but not precise

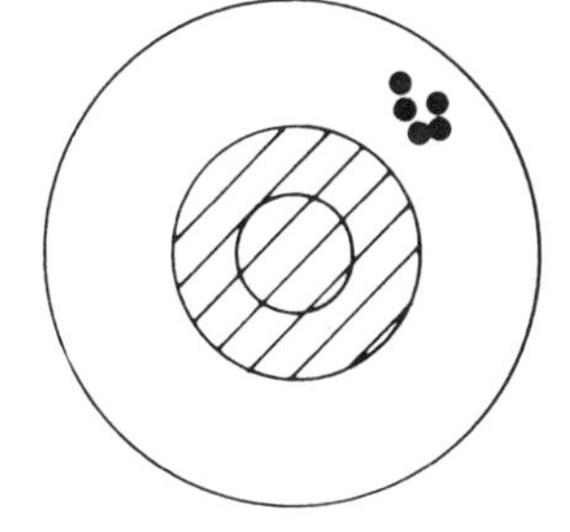

B. Precise but not accurate

▶ ***Figure 5.17*** Accuracy versus Precision

Proximity Sensors

Proximity and touch sensors of various kinds are widely used. A few of the many kinds are shown in Figures 5.18 through 5.21. They may be either mounted on the robot arm to sense targets or they may be mounted near the target to "home" the end effector or tool into the target. Knowing exactly where the sensors are mounted relative to the robot's position allows precise and accurate positioning of the tool through proper feedback servo control. The measuring sensor must be able to measure to sufficient precision for the task at hand. A sonic sensor (Figure 5.18) that is precise to 0.1 foot cannot position an arm to 0.01 foot. A good rule of thumb is that the sensor should measure one order of magnitude better (10 times better) than the measurement required. For example a LVDT (Figure 5.19) precise to 0.0001 inch can be used to position a tool to 0.001 inch.

Unless the measuring probe disturbs or damages the surface of the part being measured, it is usually better to use contact measuring devices. For fragile surfaces, noncontact probes are used. One type is the capacitance probe shown in Figure 5.20. Noncontact probes are also used to prevent collisions and for other safety considerations. Sonic ranging devices are particularly well suited for this application.

Tactile Sensors

Tactile sensors are now available with reasonable definitional and pressure resolution abilities. Many pressure-sensing elements are mounted

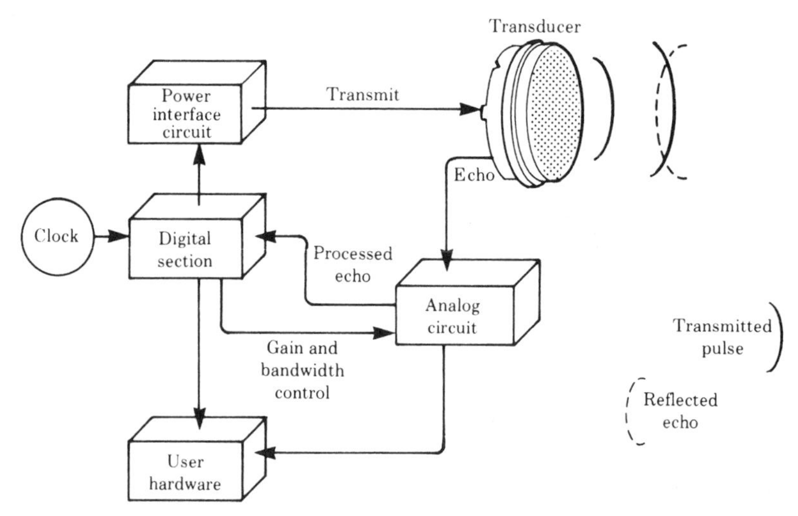

▶ ***Figure 5.18*** Sonic Sensor (Courtesy of Polaroid Corp.)

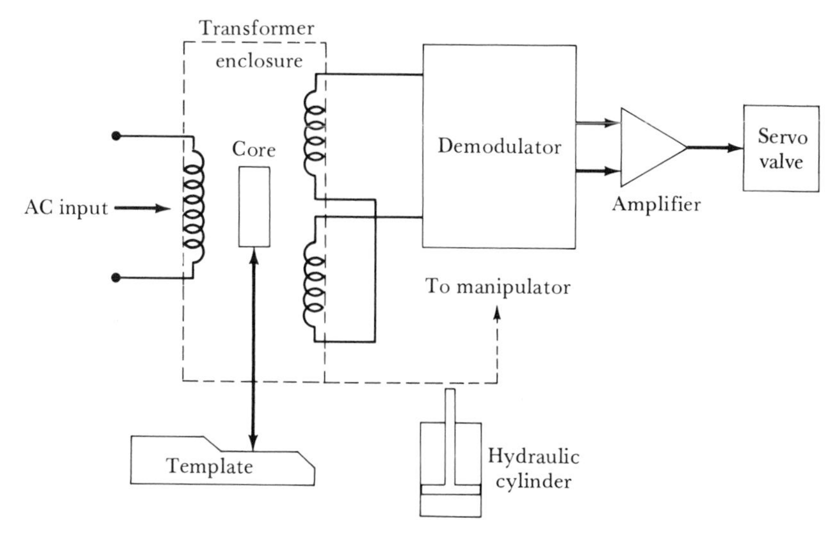

▶ ***Figure 5.19*** LVDT (Malcolm, *Robotics*)

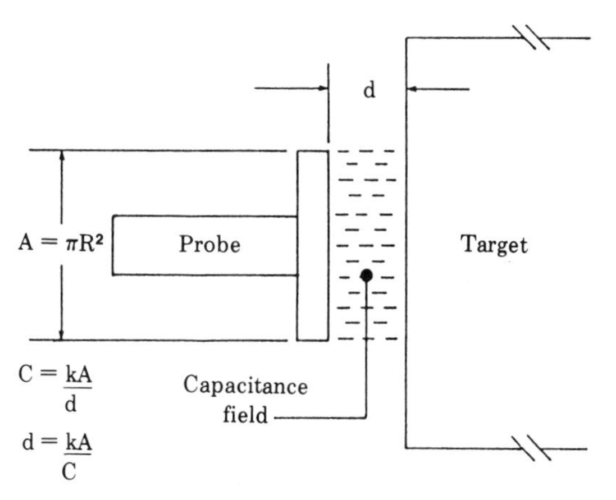

▶ ***Figure 5.20*** Noncontact Capacitance Sensor

in a rubber pad, and this pad can be glued to the gripper or placed on a surface on which the part rests. Figure 5.21 shows how this might be done. It is possible to resolve shapes such as cubes and cylinders on end, pressure gradients such as flats and bulges of the same outline, differences of mass of identical sizes and shapes, as well as location of the object on the pad. Applications are still being sought, but promising areas are handling fragile objects and picking up parts that are casually placed on the pad.

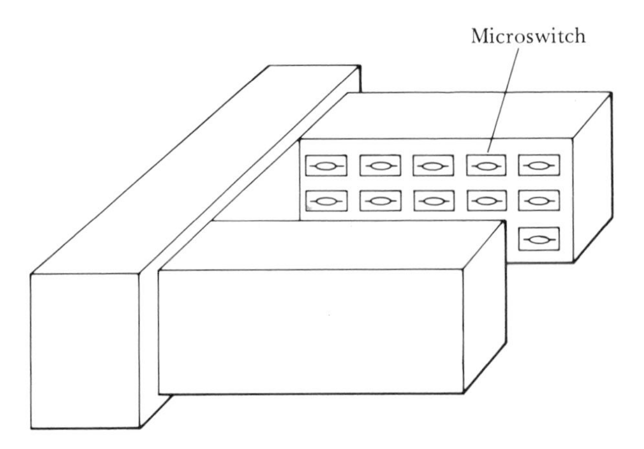

▶ ***Figure 5.21*** Tactile Sensor on a Gripper (Malcolm, *Robotics*)

Visual Sensors

Figure 5.22 shows a vision system measuring a part. Visual sensors have progressed to where not only reasonable-sized matrixes are available, with 320 × 240 points being common (76,800), but 64 and more shades of gray are available for each point. The images can be processed at the rate of 30 images per second. (See if you can figure out how many bits this is per second.) Generating the bits is not the problem. Doing something with them is. Usually what we want to know is, "Is this the right part?" or "Is it in the right place?" If yes, do A; else do B.

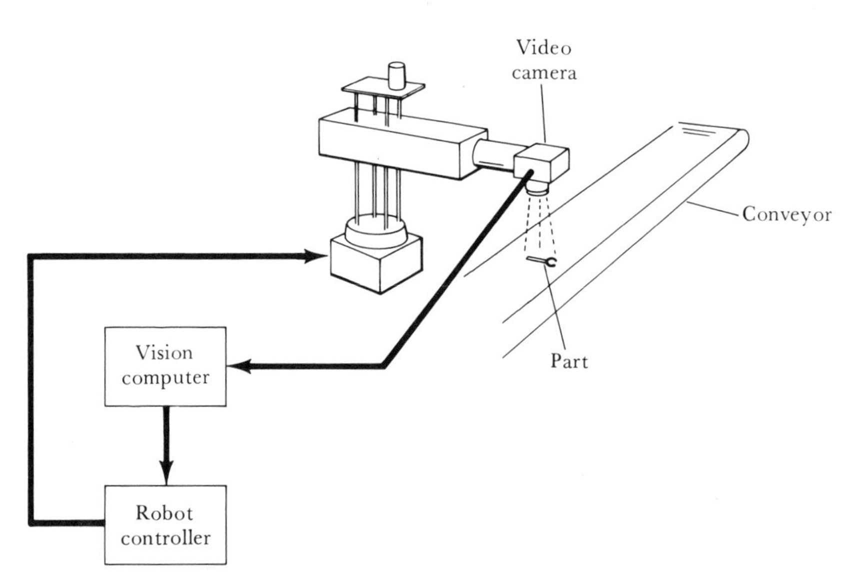

▶ ***Figure 5.22*** Vision System Measuring a Part (Malcolm, *Robotics*)

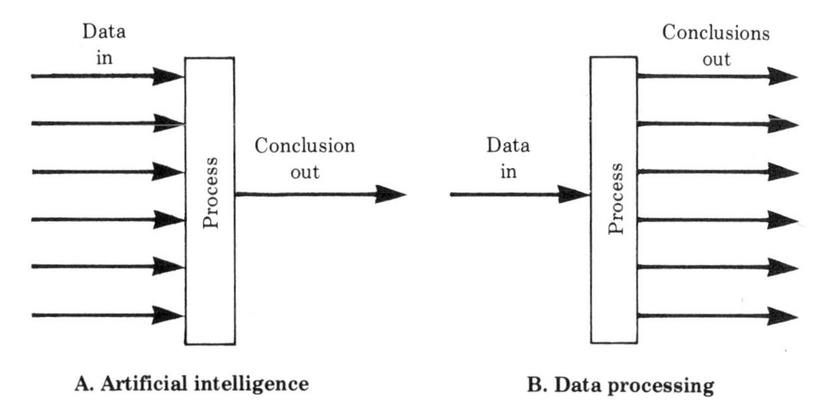

► ***Figure 5.23*** Artificial Intelligence versus Data Processing

The resultant situation is a classic problem in artificial intelligence: we have a great deal of data or input to contend with, but are looking for little in the way of output (e.g., simple answers). The typical computer takes in, comparatively, a small amount of data but provides extensive information output.

Figure 5.23 shows the difference between artificial intelligence and data processing. Processor and software technology is behind the data-gathering capability of the vision systems now on the market, but the industry is moving rapidly. Most vision systems will compare what they are looking at with an image stored in memory after processing the visual image for position and attitude constraints. There are, however, still a great many limitations. A correct label on a bottle on a conveyor line will be rejected if the bottle has fallen over prior to reading, even if the bottle is straightened later. Thus it is sometimes necessary to design the system around the equipment instead of the other way around. As with all proximity sensors, the robot needs to be prepositioned to come within the measuring range of the sensor. The sensor must be aligned or referenced to a known target. Usually, internal feedback devices provide an error signal to get the arm to an area where the sensor signal takes over to fine-tune the operation.

End Effectors

The robot arm, controls, and all the associated parts of the robot are for naught if the robot cannot do anything. The factory manager has definite ideas as to what needs to be done. The first decision is whether the machine tool with tool holder, which is all a robot is, needs to be programmable and needs to have a manipulator arm. The user then decides which robot is going to do the best job of holding and using the tool. In robotics the correct term is end effector, as the concept of a tool holder might be limited to something that only holds cutters or the like. The robot, then, is only as good as the design of the end effector. The design of end effectors is more of an art form than a science.

► ***Location of the End Effector.*** On some robots the end effector is an integral part of the manipulator or arm. Certain companys that specialize in specific industries provide the robot, manipulator, and end effector in a monolithic design. These robots are dedicated to a specific task and limited to a specific size and load range. They are usually less expensive than robots with interchangeable end effectors. However, most makers of robots do not supply the end effector because it is a "goes-onto." The end effector goes onto the manipulator. The job of the robot then is to get the end effector to the right place at the right time. Although it is more expensive than the built-in end effector, the separate end effector allows the user to make changes in both the function and load and size range of the robot. Because the end effector is a separate part, power and control interfaces must be provided at the end of the manipulator. Of the thousands of possible designs, two concepts are shown in Figure 5.24, an inside and an outside gripper.

► ***Design Considerations.*** When specifying a robot, consideration must be given to the size and mass of the end effector. If the end of the end effector sticks out a foot, this foot must be added to the manipulator end while doing the programming to ensure that the effector end is located properly. Also important is to make sure that the mass of the end effector does not overpower the robot arm. A mass that is too great may cause the end effector to whip around, resulting in damage, or the arm may sag, thus creating an error in position.

Frequently, end effectors are mounted with breakable bolts so that, upon collision, the effector will break away from the manipulator and the accident will not destroy the robot. The concept is similar to a motor shear pin. Usually, an electric circuit is activated in a collision situation to sound an alarm and halt the manipulator.

Most end effectors are designed to grip something. Gripper design is an outgrowth of holding fixture design. Because parts come in all sizes and shapes, the gripper must be designed to accommodate the particular part or class of parts. The gripper is required to take hold of the part and move it to some other location. The pressure of the grasp must be strong enough to overcome gravitational and dynamic forces, yet it must not be so strong as to cause damage to the part in question. Usually, gripper fingers or hands are either spring loaded or employ some type of compliant padding. On very fragile parts, pressure transducers in the gripper fingers are employed to provide feedback information. The contact area of the gripper is important because larger gripper areas can distribute the load and forces. Sometimes the gripper is used

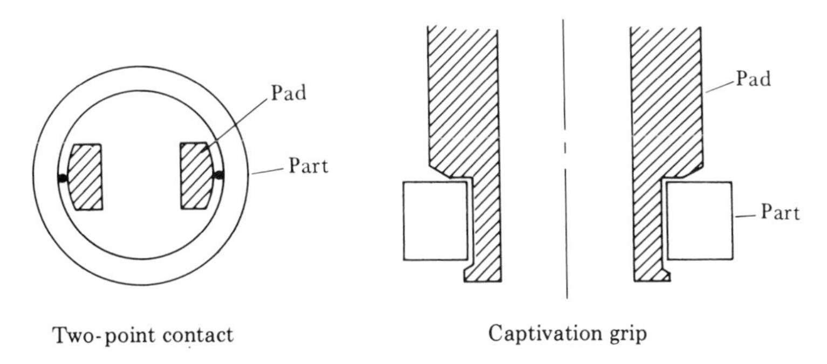

A. Inside diameter gripper

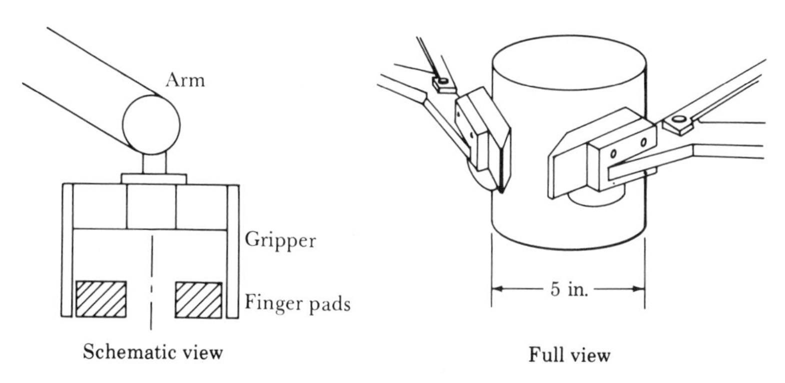

B. Outside diameter gripper

► ***Figure 5.24*** Typical Inside and Outside Grippers (Malcolm, *Robotics*)

to reorientate the part as well as move it. In this situation, due care must be given to the center of gravity of the part in relation to the center of contact of the gripper so that the part does not slip. This concept is shown in Figure 5.25. All kinds of drives are used in grippers, but mechanical gear or belt trains and pneumatic cylinders are the most common. Some parts are better handled by grasping them on the inside and these grippers clamp by opening. We frequently find two grippers on the same manipulator. In this way, the part can be passed from one gripper to another to ensure proper orientation and to prevent the gripper finger from interfering with the next operation.

► ***Vacuum Grippers.*** Vacuum grippers (Figure 5.26) are widely used in package pallet stacking and other applications. Here, a suction cup is placed on the part and a vacuum is pulled. To lift large weights, multiple cups of wide area are employed. The parts are usually configured with flat and smooth surfaces so that the vacuum will be effective.

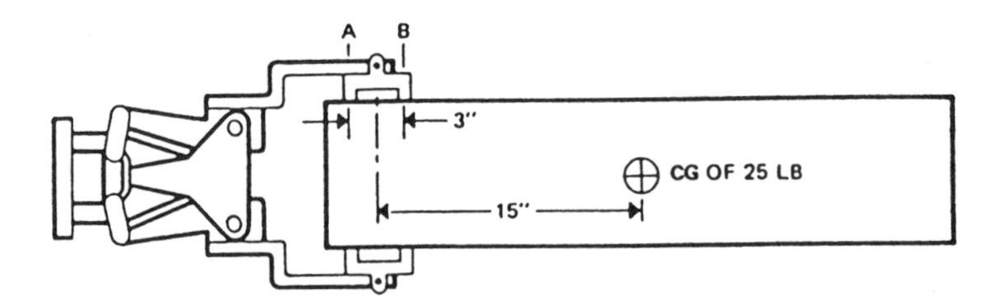

This simplifies into the following force diagram:

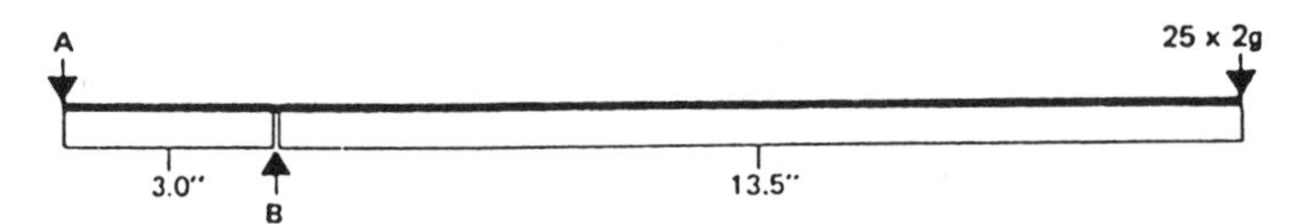

force at point A =

$$\frac{25 \times 2}{3} \times 13.5 = 225 \text{ lb}$$

force at point B =

$$225 + (25 \times 2) = 275 \text{ lb}$$

▶ ***Figure 5.25*** Center of Gravity Effects on Gripper (Engleberger, *Robotics in Practice*)

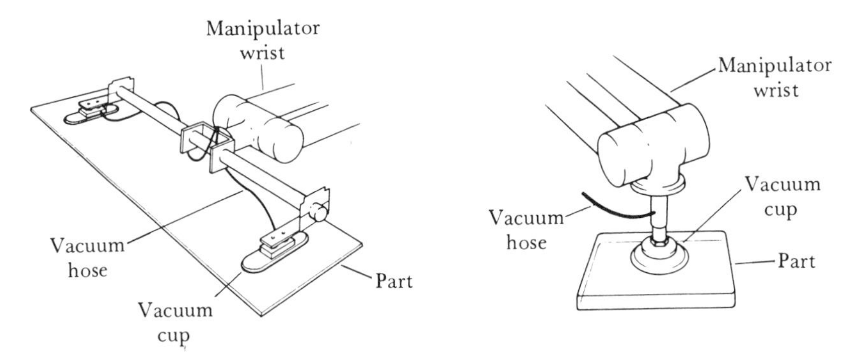

A. Dual gripper **B. Single gripper**

▶ ***Figure 5.26*** Typical Vacuum Grippers (Malcolm, *Robotics*)

One advantage of the vacuum gripper is that the cups are rubber and do not mar the surface of the workpiece. Similar to vacuum grippers are magnetic grippers. These only work on ferrous materials, however.

▶ ***Tool Holders.*** Any kind of tool holder can be attached to the end of the manipulator arm depending on what needs to be done. Cutters and milling bits are common tools used for shaping operations. A typical end of the arm tool holder is shown in Figure 5.27. We also find exotic

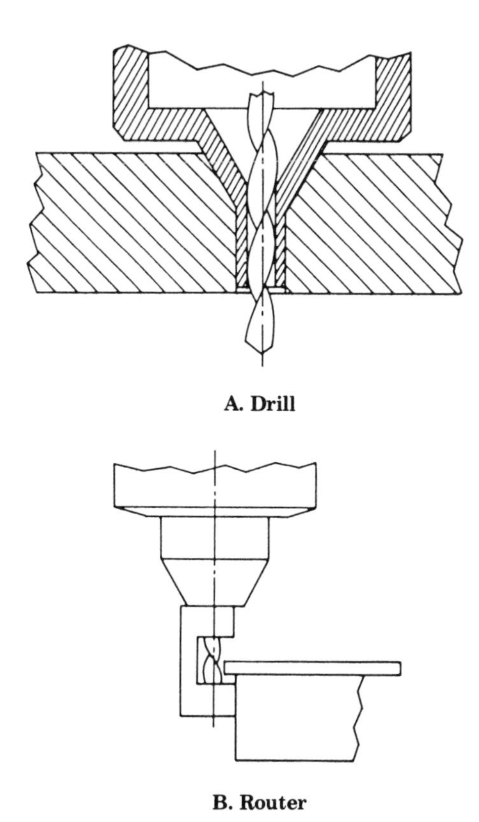

▶ ***Figure 5.27*** Typical End of Arm Tooling (Malcolm, *Robotics*)

tooling used in robotics, such as laser cutters, water-jet cutters, electro-machining, and deposition electrodes, as well as specialized welding, insertion, and assembly tools.

5.5 Applications of Robots

As mentioned earlier, robots are used in undesirable locations to do tasks that are not attractive to humans. Some of the uses to which robots have been put are as follows:

- Die casting
- Spot welding
- Arc welding
- Investment casting
- Forging
- Press work

- Spray painting
- Plastic molding
- Foundry practice
- Machine-tool loading
- Heat treatment
- Deburring metal parts
- Palletizing
- Brick manufacture
- Glass making
- Material transfer
- Inspection
- Testing
- Assembly

A very early application of robots was in the nuclear industry for handling radioactive isotopes. These early machines were manipulator arms that were in some cases mechanically cam operated. The cams were removable and hence it was possible to reprogram the movement. New uses for robots are being found everyday.

▶ ***Need to Be Economical.*** Robots are used only where it is economically attractive to do so. In any process, we must establish the rate of production and the quality level we wish to achieve. In the welding application cited earlier, robots are attractive because they can produce consistently high quality, uniform welds at the fastest possible rate. A human can produce welds of high quality only intermittently and at a rate that is only sometimes the fastest possible. If we could find a human that could produce as well as a robot, and the "capitalized" value of the human were less than for the robot, we would not use robots. With most tasks in industry it is still cheaper and better to use humans, which is one reason why the market growth for robots has been disappointing.

▶ ***Robots Are Used in Unattractive Locations.*** In addition to hazardous jobs and physically unattractive jobs that are applicable for robots, there are jobs that are boring. Few people are consistently productive in boring jobs. A friend listens to Mozart and other interesting things while she is doing data entry for the IRS. She claims she has no conscious knowledge of what she is entering into the computer and her reflexes are purely automatic. She is rated as a very productive and accurate worker. The rest of us, however, are usually painfully conscious when we are doing boring tasks and as a result our productivity suffers. For this reason, robots are used in repetitive jobs like loading, unloading, and component insertion. Photo 5.4 shows a Milacron robot loading and unloading an injection-molding machine.

▶ ***Photo 5.4*** Load/Unload Robot Application (Courtesy of Industrial Robot Division, Cincinnati Milacron)

▶ ***Need to Define Tasks.*** A task that can be explicitly defined in terms of motions and sequences has a better chance of being automated than one that cannot. In addition, if it is possible to define the contingencies of the process, then automation is more feasible. The more motions, sequences, and contingencies we have, the larger the computer hardware and the more complex the software. An opponent of the Star Wars concept, a computer expert at MIT, claims that Star Wars is not feasible, even if we can develop the hardware, because we will never be able to define all the contingencies. Programming for the space shuttle is in the million-program-line range because we can predict the rotation of the earth and such factors. But we don't know what our enemies will do in a space war yet and have no good way to test any possible theories. The amount of programming is still open ended. The point is that programming is very costly. If our process is such that we need to reprogram every time we run the robot, we will use humans. If the process is such that we require long runs of a repetitive nature, we will opt for hard automation. Robots fit somewhere between these two extremes. In a similar vein, robots are used where the program requirements are reasonable.

▶ ***Proper Applications Are Required.*** Robots are frequently found in applications in which they do not belong. Part of the reason is that robots have been oversold to industry management. The promise is that they never get sick and don't drink coffee. Sickness, coffee breaks, employee turnover, and the like, are all measurable expenses, as are replacing component parts, preventive maintenance, operator retraining, and reprogramming.

5.6 Summary

Robots are a useful illustrative tool for CIM because they employ a great many of the components found in automation equipment and other robotic devices, and they are programmable by virtue of having their own computer. Robots are a specific machine tool used in hazardous, boring, and other unattractive locations. End effectors are an important part of robot design.

Servos are important in all CIM equipment including robots. A servo system provides much better resolution or precision than can be accomplished without it. In robots, servos are most often used to provide position and velocity information to the control circuits in the computer.

5.7 Exercises

1. What is the RIA definition of a robot?
2. What are the key characteristics of a robot?
3. How do we classify robots?
4. What are some advantages of using a robot?
5. Name the two different types of arm geometry.
6. What is the difference between precision and accuracy?
7. What do we mean by degrees of freedom?
8. Of what class is a robot a species?
9. Describe some of the characteristics of a power supply.
10. What is a motherboard?
11. What is the function of a control panel?
12. What are some of the controls you might find on a control panel?
13. What is meant by slippage and backlash?

14. What are some of the ways of getting around backlash?

15. What are some of the problems associated with control loops?

16. Describe some of the consequences of using closed-loop control systems.

17. Discuss noncontact sensors and LVDTs in relation to precision.

18. When we talk about an end effector, what do we mean?

19. Why do we need to supply interfaces to the end effector?

20. What two design characteristics must be considered when selecting a robot?

21. What do most end effectors do?

22. Why do we need to be concerned with the center of gravity of the part in relation to the end effector or gripper?

23. Where are robots generally used and why are they finding popularity in these applications?

24. Why do we sometimes find robots where they don't belong?

6

CIM Units: Material Handling

6.1 Introduction

Overall Material Flow

In earlier chapters we spoke about islands of automation and manufacturing cells. In a typical plant we are likely to find an area where raw material is received. After the goods have been checked for correctness, they are sent to a storage area where they are kept until needed. As part of the checking, the goods may be bar coded and read automatically or the part numbers and descriptions may be entered into the computer manually. In the storage area, the goods are placed in bins or on racks in an orderly fashion where each individual item can be found quickly and accounted for. Material storage areas are highly automated. When a raw material is called for to be worked on or processed, it must physically move to the machine center, cell, or island where the fabrication operation takes place. Concurrent with the movement of the material to the fabrication center, other parts must arrive before any work can begin. For the most part, the other parts are tools, jigs, and fixtures. How the materials and "other things" get there is the subject of this chapter.

Machine Tools and Tooling

To help understand the process better, we will take the time to discuss some terms. A machine tool is a power-driven device that holds the tool that does the work. Typical machine tools include lathes, milling machines, broaching machines, and grinders. The tool is the part that actually contacts the metal or other material on which work is being done. Drill bits, cutter heads, and grinding wheels fit into this category. Usually, the tool is held by an intermediate piece, called the tool holder, such as a chuck or collet, which screws into or attaches onto the machine tool. The tools for a particular machine come in different sizes and must be compatible with the tool holder for that class of machines. For example, router cutting heads are different shapes, but all the shanks are the same diameter for gripping by the collet.

Fixtures and jigs are material-holding devices, and some are standard, such as a vise or clamp; but many are especially designed and fabricated for a particular function. Figure 6.1 shows how a machine tool with tooling and the workpiece might be configured. A drill press stand is used as the model. Designing jigs and fixtures requires imagination and skill. Before a job can be run, all the material, tools, and fixtures must be placed on the machine, and the machine must be adjusted with the right settings for speed, feed rates, coolant flow and many other details. All these activities come under the heading of setup.

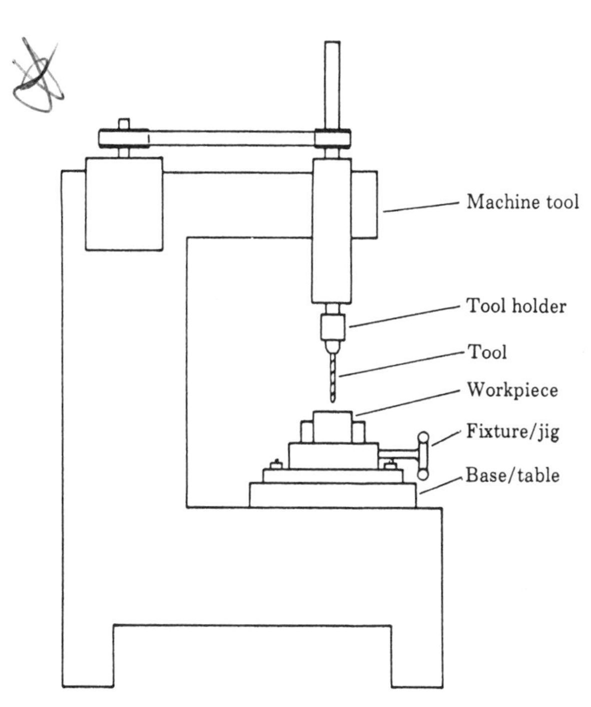

▶ ***Figure 6.1*** Machine Tool, Tooling, and Work

▶ ***Getting Parts to the Machine Is the Key.*** Once the machine has been set up, the fabrication cycle can begin and a part can be worked on. Typically, the time it takes to make the part, which is the actual cutting or grinding operation, is very short in comparison to the amount of time it takes to set up the machine to make the first part. Since the setup charges must be prorated onto the cost of the part, a certain number of parts must be fabricated to make the setup and run worthwhile. The parts that are made but not used in the next operation end up as work-in-process inventory, and keeping a part on hand for 6 months or longer is not unheard of.

Now we need to look at how the materials, tools, and fixtures get to the machine tool. The choice of conveyance method depends mainly on the lot size of the run. The bigger the lot size is the more hard automation we can use and the less flexibility we have in part and process design.

The people who design and manage a plant usually know what machine tool they need to make a part. Less certain is the method to use to get the parts into and out of the machine tool itself. Although the idea of a conveyor is simple, designing a conveyor to do what is wanted is not. Neophytes are surprised to find that conveying and transporting equipment is very expensive, and unless great care is taken the system ends up as a dedicated piece of equipment. It is a fairly simple matter to move a drill press 50 feet to the right, but not a walking beam assembly line (Figure 6.2). Preparation, transporting, and conveyance of materials, parts, tools, and fixtures, although unglamorous, frequently amount to

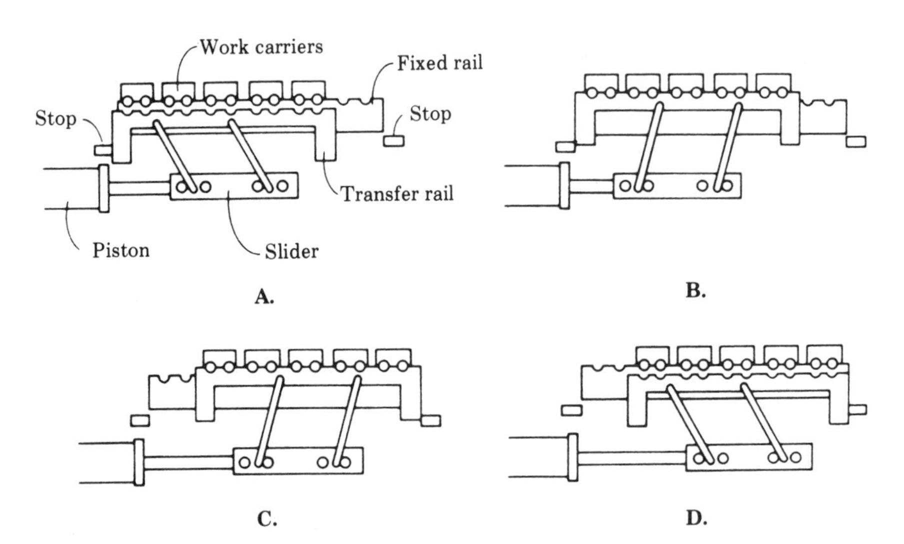

▶ ***Figure 6.2*** Walking Beam Transfer System (Boothroyd, *Automatic Assembly*)

more than 50% of the capital equipment in a manufacturing facility. And, if a change is required, it is sometimes cheaper to start from scratch and build a new plant than to move machine tools around.

6.2 Automation Equipment

► ***For the Modern Factory We Need Programmable Equipment.*** The easiest challenge is to make equipment to transport the same part to the same place forever. Early automotive and appliance lines were set up to do just that. Unfortunately, the consumer, the person who pays for all of this, now wants different configurations and options. To satisfy these requirements, it is necessary to build versatility and flexibility into the system. To plan, implement, and control the flexibility and versatility, we rely on computers to do much of the job. They are good at keeping track of things. Much of the automation equipment found in a CIM plant uses the same principles and mechanics found on older assembly lines. What is different is that the equipment needs to be able to move an assortment of parts to various locations. In addition to being versatile, the equipment must be programmable and must be able to fit into the computer network.

6.3 Loaders and Unloaders

In this book, a part or piece is anything being moved around, whether it be raw material stock, tools, jigs, and fixtures, or what have you. It is something that is transportable. At the machine tool site we need to have a machine to load and unload the part into the machine. In plants that are not automated, this is one of the main functions of the operator. Another principal function of the operator is to gauge the parts and adjust the machine tool so that the correct amount of material is removed from the workpiece by the tool. A simple pick and place mechanism takes a part and puts it in the correct position.

Figure 6.3 shows a robot removing a part from one carrier and placing it on another. In the warehouse a robot is used to remove a part from a bin or shelf and place the part on a conveyor. At the machine tool end, the pick and place removes the part from the conveyor or station and places it for loading into the machine. Precision depends on the mechanical condition of the mechanism. Too much play or slip and the part ends up in the wrong place. A gripper or clamp is used as well as simple hooks in some applications. Robots are widely used for loading and unloading.

Sometimes the part needs to be turned over or reoriented. The part arrives to the station facing north–south and the machine tool is east–west. The machinery that changes the attitude of the part is called

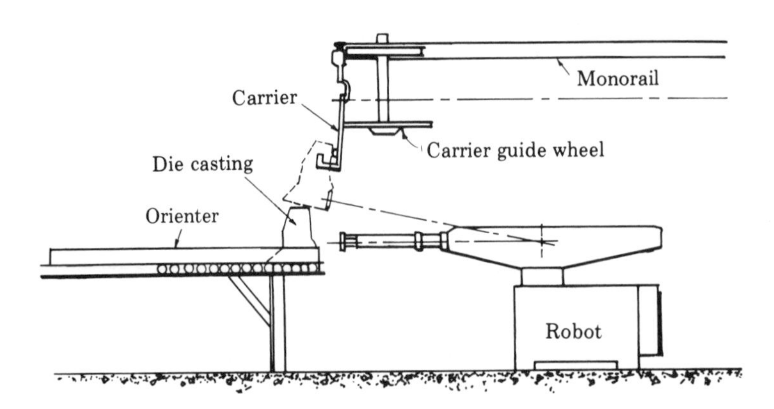

▶ ***Figure 6.3*** Machine Transfer Work Cell (Malcolm, *Robotics*)

a positioner. One type of machine does nothing but turn the part over. Others lay the part on its side. Often a gripper is used, which is either mounted on a table or sometimes a wrist. In some cases a swivel action is required. Some part-orientation machinery has many degrees of motion. When made programmable, they become robots, and robots are frequently used for these simple tasks. Other applications require the part to be placed on a table and indexed or rotated so many degrees so that it may be picked up, worked on, replaced, and then indexed to the next station for further operations. In some machines the various functions are combined. Usually, the machine is designed for the particular part or process. Various assembly configurations are shown in Figure 6.4.

▶ ***Automatic Gauging Is Frequently Added.*** An aid to positioning frequently found on an automated line is an automatic gauge or series of gauges. A sensor, usually an LVDT, is positioned against the part being fabricated and is referenced to a fixed point in space so that dimensional measurements can be taken. Output from the sensor can be read out in a number of ways, from the operator looking at the readout or dial to a direct computer interface with the sensor. In positional automatic gauging, positioning of the sensor to high precision is essential. Other automatic gauging will sense the parts for correct dimension and sort the rejects from the acceptable ones.

▶ ***Most Units Are Custom Designed.*** Figures 6.5, 6.6, and 6.7 show three ingenious examples out of the hundreds of mechanisms that can be used for transfer systems in both fabrication and assembly. Large pieces such as engines are fixed to a work carrier, and the work carriers are moved around. The loads and speeds needed by the loaders, unloaders, and positioners determine how they are driven. Drives may include simple mechanical power trains (e.g., gears, belts and chains, cams and

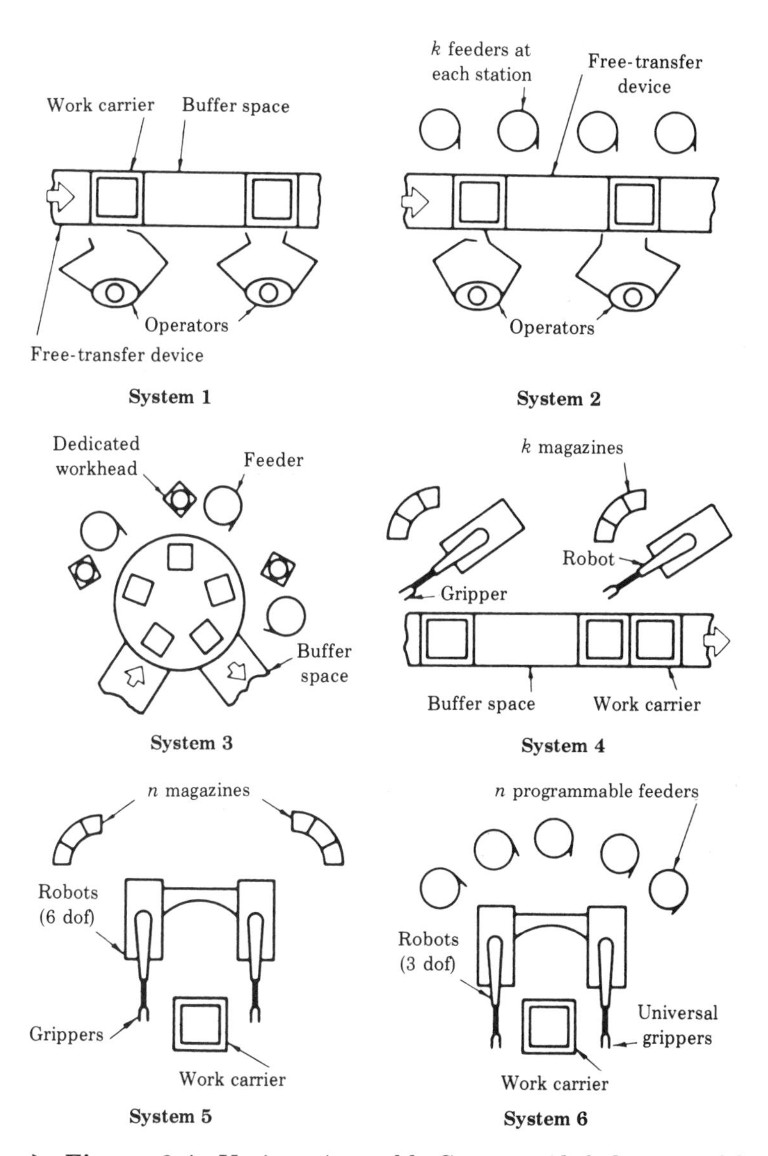

▶ ***Figure 6.4*** Various Assembly Systems (dof, degrees of freedom) (Boothroyd, *Automatic Assembly*)

cogs, etc.) or pneumatics and hydraulics. Switches of every conceivable variety are employed to actuate the mechanisms, and these switches may be actuated by sensors or may be computer controlled. Older devices used relays, but these have been replaced with dedicated programmable controllers, which are usually less costly and more reliable. Reprogrammability is not necessarily important as many machines are usually dedicated anyway.

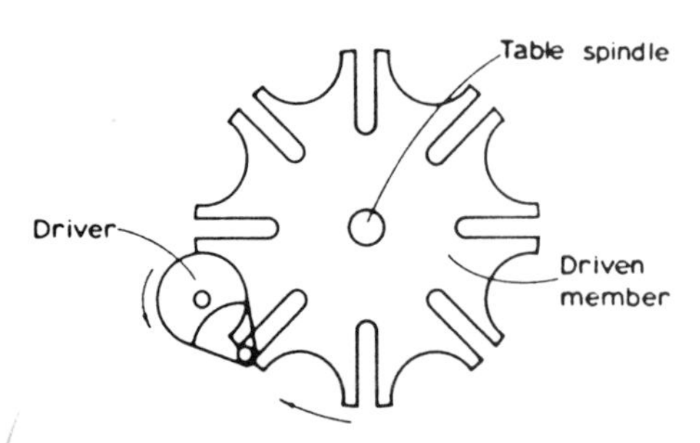

▶ ***Figure 6.5*** Geneva Mechanism (Boothroyd, *Automatic Assembly*)

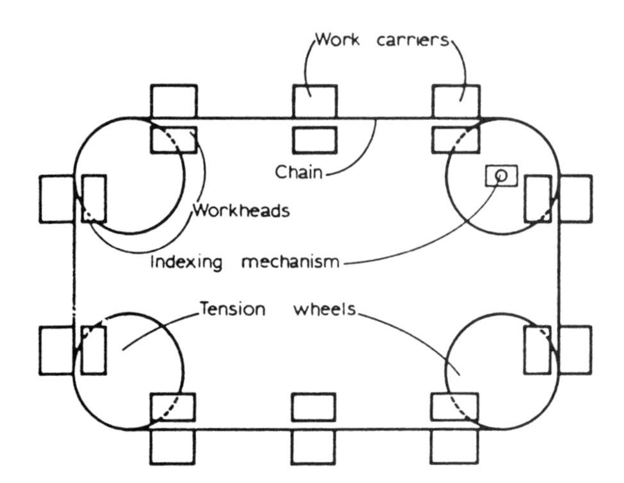

▶ ***Figure 6.6*** Chain-Driven Transfer System (Boothroyd, *Automatic Assembly*)

There is some standardization in design with transfer mechanisms, but not much. Usually, each machine tool is different, as is each specific loading and unloading task. There is an industry of highly specialized manufacturers of this type of equipment and the engineering content in the finished mechanism is usually quite high. However, some companies, such as Weldun Automation Products, offer modular units such as those shown in Photos 6.1 and 6.2.

6.4 Accumulators

▶ ***We Need Temporary Storage to Balance the Line.*** Some machine tools can make parts faster than others, so it becomes important for the plant engineering group to balance the line. It is very difficult to

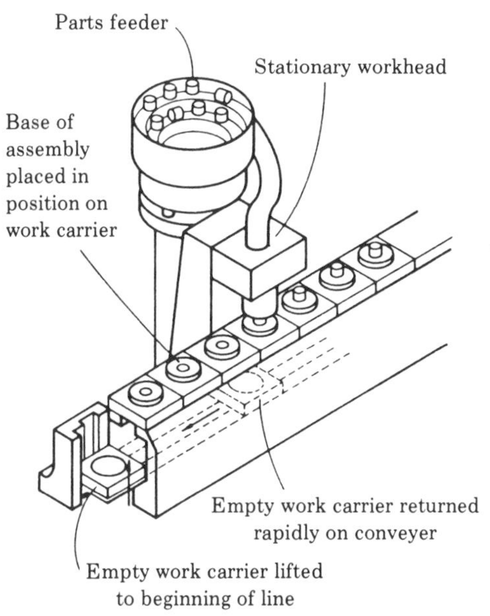

The machine is shown with shunting work carriers returned in the vertical plane.

► ***Figure 6.7*** In-Line Transfer Machine (Boothroyd, *Automatic Assembly*)

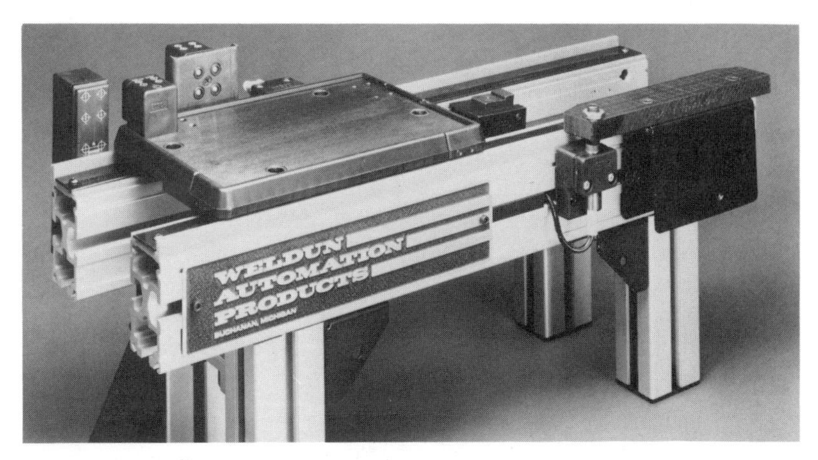

► ***Photo 6.1*** Elements of Modular Transport (Conveyor) System (Courtesy of Weldun Automation Products/Bosch Group)

adjust conditions so that a part is always there just when needed. Therefore, accumulators are used in the operations sequence to take into account the differences in machining operations. Accumulators are

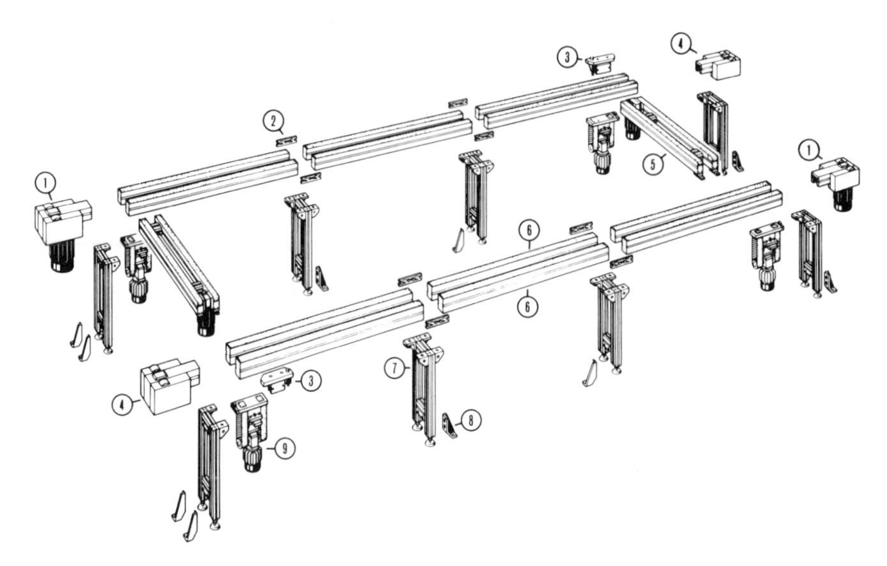

► ***Photo 6.2*** Coding Information System Offers Optimum Flexibility in Traffic Management (Courtesy of Weldun Automation Products/Bosch Group)

not to be confused with work-in-process (WIP) storage. WIP storage implies that a batch of parts are made and then held until they can be processed further. An accumulator is required to smooth out the operations. Let's take an example. A certain machine operates at an optimum rate depending on the amount of material to be removed, the hardness of material, and other physical factors. If the machine operates too slowly or too quickly, bad parts will be produced. In a balanced operation, parts arrive at the machine at a certain rate and can be removed from the machine at the same rate. For the machine to make good parts, it has to run at optimum speed, receiving parts at its designed rate, which may be faster than the infeed rate. The machine runs for a while, drawing down parts from the accumulator, until the machine has no more parts to work on; it then stops until the accumulator has been replenished. Usually, the cycle time is on the order of seconds, or perhaps minutes at most. In a plant where they make foam mattresses, they run the foam machine for about an hour each day and accumulate enough finished foam for a full day's fabrication. A large crane removes the buns from the line for cooling and stores them in a large adjacent warehouse until needed. The reason the machine is not run continuously is that if a foam machine were used that employed a smaller throughput rate, they would end up with mattresses that were 40 inches long. There isn't much of a market for 40-inch mattresses.

▶ ***Accumulators Are Also Custom Designed.*** Like loaders and unloaders, accumulators come in all sizes and shapes depending on the need. Parts that will be used in an insertion machine are frequently accumulated in a helically configured ramp. This type of accumulator is frequently called a blue steel accumulator from the type of spring steel used to make the spirals and helixes. Typical parts might be automotive valve lifters and wrist pins. Figure 6.8 shows two types of accumulators. Parts for insertion into printed circuit boards are frequently held in magazines or clips like ammunition carriers. Sometimes the parts are held on the conveyor by a flapper, such as on a bottling line. Very small parts are frequently loaded into a vibratory feeder in batch fashion, where they are jiggled and cajoled as they move up a ramp until they emerge right side up. Along the way the upside down parts fall back into the bin for recycle. Figure 6.9 shows one such device.

Stackers are used extensively in stamping and forge operations where a pile of material is off loaded one at a time into the machine.

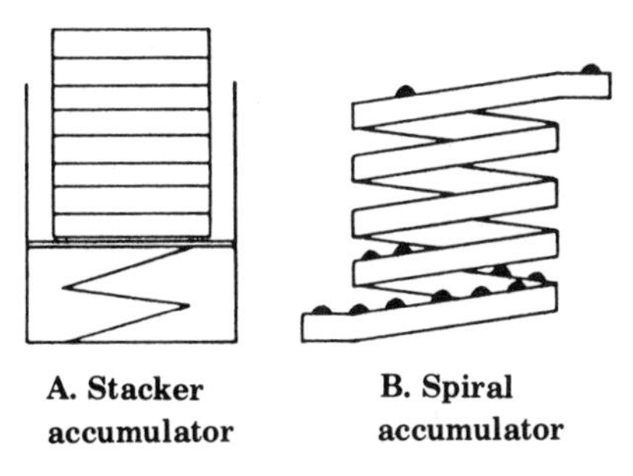

▶ ***Figure 6.8*** Two Common Accumulators

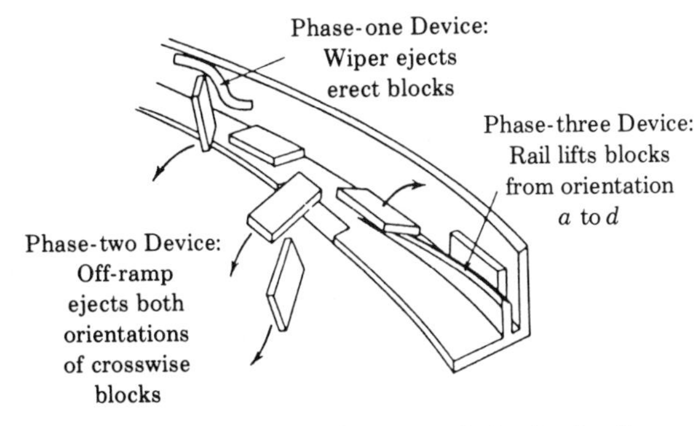

Shown are the orientation and rejection device for the vibratory bowl feeding of rectangular objects.

▶ ***Figure 6.9*** Bowl Feeding (Asfahl, *Robots and Manufacturing Automation*)

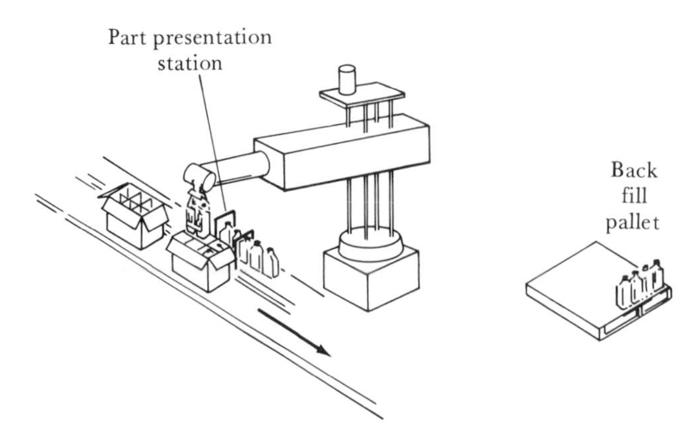

▶ ***Figure 6.10*** Robot Palletizing Bottles into a Carton (Malcolm, *Robotics*)

Unstackers take the material from the machine and place it in a position for transport to somewhere else. Palletizing, which is a form of stacking, is a major application for robots due to their flexibility and reprogrammability. Once the part is assembled, it needs to be packaged into boxes and the boxes placed on pallets for shipping. Figure 6.10 shows a typical unloading and palletizing operation.

6.5 Conveyors

Conveyors come in all sizes and shapes. Conveyors move material from one place to another, although on one-product assembly lines they are also used for work-in-process storage. Monorail conveyors can either be operated by pushing a trolley along the rail or by employing a driven chain to move the trolley along. See Photo 6.3, which shows a conveyor moving cars in a Fiat plant. Attached to the trolley is a hook or some other device to hold the parts, for example a bucket or gripper. Some conveyors use two rails and the parts can either be suspended from the conveyor or placed on top of cross members attached to the moving chains at each side of the conveyor. Large beams carry heavy parts, but small parts are usually placed on a wire mesh or belt suspended between the two rails. Things in boxes frequently are placed on roller wheels or balls mounted across the conveyor, which move by inertia or gravity. A special configuration called a race track, which is generally oval shaped, uses cars that are pulled along by a chain. It is similar to what you find in an amusement park or car wash.

Conveyor rails are fixed or mounted to the structure, and the path is fixed. Switches are used to move material to other conveyors or

▶ ***Photo 6.3*** Assembly Conveyor (Courtesy of Texas Instruments Incorporated)

off the line to work stations. Removal methods range from manual to microprocessor-controlled mechanisms such as pneumatic flappers and switching points. Frequently, a conveyor crosses an aisle where either a break in the conveyor is required or up and down ramps are employed. Where space is tight, it is sometimes necessary to incorporate a lift. Photo 6.4 shows an automated conveyor system in John Deere's highly advanced CIM factory.

6.6 Storage

▶ ***We Will Always Need Some Storage.*** As goods and supplies come into the factory, they must be placed into inventory until needed. In keeping with the just-in-time philosophy, it would be ideal if the vendor delivered the supplies to arrive at the fabrication or assembly point as the operator were starting the machine. Many companies are integrating their vendors with their requirement planning to accomplish these ends. Since we live in an imperfect world, we need to plan on shipments being late, and to take advantage of purchasing discounts, we sometimes have to take delivery in excess of our immediate needs. Material storage is an inescapable fact of life. As the complexity of manufactured products increases and as the number of options needed increases due to competitive factors, inventory control and storage needs increase geometrically. It is not uncommon to have 100,000 part numbers and millions of parts on hand in

► ***Photo 6.4*** Automated Conveyor System (Courtesy of John Deere & Company)

even a modest operation. It simply is not possible to do it any other way except by computer.

Although the computer has been used to keep track of inventory for many years, the use of computer-driven equipment and machines to actually effect the storage of parts has not been in widespread use until fairly recently. The name for this is automatic storage/automatic retrieval (AS/AR). In the factory-without-lights, material is delivered to the door, checked, entered into the inventory control system through readers, physically transferred to the storage area by machine, and stored in the appropriate location without ever having been touched by a human. When a shop order reaches the storage area, the proper quantities of each subassembly or part are removed from their individual locations, conglomerated, and transported to the desired location, again with automated machinery. The warehouse terminology for the conglomeration process is called picking. If we take the individual parts or subassemblies and package them into assembly units, this is called kitting.

► ***Large Parts Are Handled Manually.*** Large and heavy parts are frequently stored in the warehouse on racks. These racks are located in bays or long aisles so that a fork lift truck can travel down the bay to the appropriate location and then raise the lift to the correct level to fetch the part in question. This is essentially a manual operation because of the cost to automate related to the return. Although it would be possible to build an automatic storage system to emulate this type of operation, the extensive hardware and power drives required would be very expensive and thus might not be justified.

▶ ***Small Parts Are Kept in Bins.*** Smaller parts can be placed in special boxes and the box loaded onto a carrier. This carrier is mounted on a track or rails and it transports the box to the proper storage location, lifts the box to the correct height, and places the entire box into its proper bin location. The person who places the parts into the box must check to see that nothing sticks up too high or the system will jam. An operator must assign the box to the proper bin location and must enter this information into the computer. When a pick is wanted, the parts list is retrieved from the computer and transferred to the stacking equipment. Transporters go to the proper bin location and retrieve the box, and the correct number of parts is removed from the bin storage box and placed into the kit box along with the other parts in the kit. The storage box is then returned to its proper bin location. Photo 6.5 shows a typical bay with a programmable picker at Whirlpool Corporation. A totally automated warehouse system is offered by Texas Instruments and is shown in Photo 6.6.

▶ ***Rely on Vendors to Repackage.*** The discerning student will have observed that there are still quite a few manual operations in this

▶ ***Photo 6.5*** Bay with Automatic Picking (Courtesy of Whirlpool Corporation)

▶ ***Photo 6.6*** Totally Automated Warehouse (Courtesy of Texas Instruments Incorporated)

process. A human must pick the correct number of parts out of the storage box and transfer them to the kit box and must enter this information into the computer. This is a potential spot for an error to occur. Placing the parts into the storage box is another error-prone operation. Always keep in mind that the main consideration in CIM is to ensure quality by the elimination of errors. A possible solution to the question of getting the right number of parts into the kit box is to have the vendor package the parts in convenient kit sizes so that whoever does the picking does not have to count the actual parts but simply removes a bag or package that contains the right number. Some companies ask their vendors to go one step further and ship the parts prepackaged and in special bins, which the user supplies to the vendor. There has also been progress in using vision systems to ensure that the correct number of parts is transferred from box to box, and this application promises to expand. Right now, however, the best solution is to get good people and motivate them properly.

Even smaller parts are stored in an oval-shaped storage device that moves around on a track or overhead monorail. Baskets suspended from the rail contain the individual parts. When a part is called for, the entire mechanism indexes around and the proper bin is brought to the front so the part can be picked. The computer control, inventory input, and other problems are similar to the bin arrangement described earlier.

▶ ***Parts on Tapes Are Used in High-Speed Assembly.*** In high-speed assembly operations, such as might be found in an electronic plant where printed circuit boards are assembled, a significant amount of the components are prepackaged to work with the automatic assembler. Resistors, diodes, and similar parts are taped together in a belt configuration very much like ammunition for a machine gun. Components such as integrated circuits are packaged into tubes or magazines that are loaded into the assembly machine. The prepackaging can be done by the assembly plant where they are used, but more often the vendor supplies the parts prepackaged in reels or magazines. However, the vendor must be certified and parts conform to quality requirements. In addition to reducing handling at the assembly location, proper prepackaging ensures that diodes are oriented correctly and ICs are not destroyed through electrostatic discharges from improperly grounded people.

6.7 Automatic Guide Vehicles

▶ ***AGVs Move Material along Set Paths.*** The automatic guide vehicle (AGV) is a fairly recent innovation. In its simplest form, it can be thought of as an operatorless fork lift truck or golf cart. Most are battery operated and require a fair amount of maintenance. Parts are conveyed onto the bed of the vehicle and the vehicle is dispatched along its way by radio. The vehicle moves along set paths in the factory. When it arrives at the proper destination, the material is off loaded onto another conveyor or at the work station. The vehicle is then dispatched to the next location or to home to await further orders. A computer controls the travel and destinations of the vehicle and can program it to follow different paths and to start and stop at appropriate times. Figure 6.11 shows a typical AGV system.

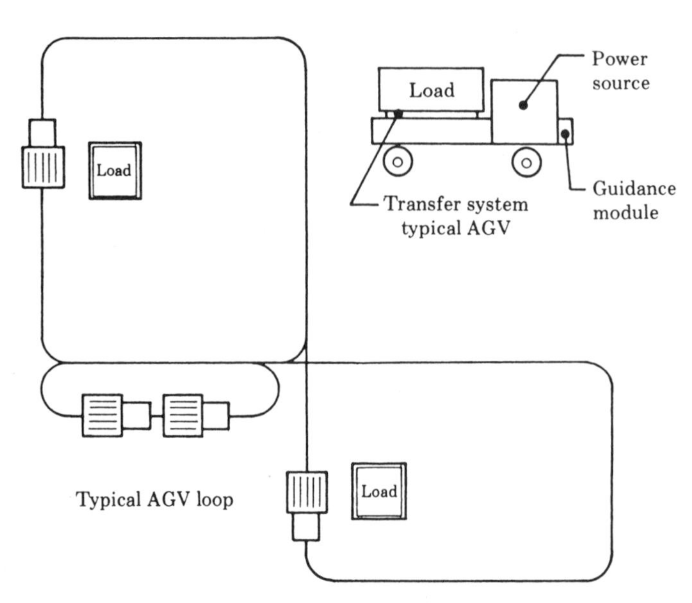

▶ ***Figure 6.11*** AGV System

▶ ***Vehicles Follow Wires or Tapes.*** Nearly all vehicles move on the factory floor on rubber wheels as opposed to moving on tracks. It is very expensive to install tracks. There are many ways to lay out the path, but the two most common use either a wire embedded in the factory floor or a reflective tape placed on the floor. In the case of the wire, a signal is generated that a sensor on the vehicle picks up. When the vehicle starts to wander away from the wire, an error signal is generated that effects the steering mechanism on the vehicle (a servo system in other words). Some systems modulate signals on the wire for vehicle commands and thus eliminate the need for two-way radio equipment. The reflective tape method uses a light source and photocells mounted on the vehicle. As long as the reflected light is at the proper level, the vehicle stays on the path. The servo controls are similar in the photo system.

▶ ***Choice Depends on Application.*** The embedded wire has the advantage of not being easily harmed. However, because it is necessary to dig a small groove in the floor, lay the wire, and cover it up again to make the floor smooth, all of which stops production, installation of embedded wire is a costly operation. If the floor has tiles, the tiles must also be removed and replaced. The tape method is obviously easier to install and certainly much more flexible when it comes to changing paths. The disadvantage is that the tape is comparatively easier to destroy than the embedded wire. Cross traffic of people and vehicles wears the tape, and spills and stains cover up the reflective surfaces. The initial cost of the wire system is higher, but the maintenance costs of the tape are higher. An important consideration, then, is how long the system will be in use before the paths are changed. In one electronics assembly factory, paths are changed frequently as the business grows.

► ***Safety Is Always a Consideration.*** The vehicles must travel slowly to avoid safety hazards. Beepers and flashing lights are employed to warn of their approach and the paths are clearly marked out on the floor. Transducers mounted on the vehicle in strategic locations sense when things or people get in the way. The vehicle is programmed to stop when objects are too close or when it hits something.

In a crowded factory environment, the tendency is to place things in the aisle for temporary storage. This is, of course, contrary to usual company policy for safety reasons, but it happens anyway. Not having the parts handy will affect productivity in a department right now, but something in the aisle may or may not cause an accident sometime in the future. Unless the infraction is truly gross, the traffic in the aisle can usually step around the offending pile. In a factory that has AGVs, things in the aisle will stop the AGV and thus affect production. The AGV aisle must always be clear.

6.8 Washers, Lubricators, and Ovens

► ***Ancillary Equipment Is Expensive.*** Sometimes relegated to the afterthought category, washers, lubricators, ovens, and other essential equipment used to ensure that the process makes acceptable parts constitute a major capital expense. Because of their size, bad results in material flow invariably occur if they are not given due consideration. Each type of equipment involves an infeed transfer and an outfeed transfer, and in this regard they are similar to other machines. Transportation of parts through the devices has conveyorlike characteristics. Care must be taken to ensure that the parts are oriented properly so that fluids, in the case of the washers and lubricators, and heat energy, in the case of ovens and furnaces, are applied properly. Robots are used to load the hooks, baskets, or whatever, and sometimes robots actually reach into the oven or washer to move parts around.

In factories where large batches of the same part are run, washers and ovens have a few simple controls. Sometimes we find programmable controllers used in these applications because cycle times are frequently important. In the case of the oven or furnace, we need to control both the amount of heat and the heat rate and duration at given levels of heat. Washers come in many sizes and can have various stages depending on how dirty the part is, the type of dirt (e.g., grease, mud, paint), and whether we need to chemically activate or passivate the surface for the next operation.

▶ ***Parts Are Frequently Sent Out for Finishing Operations.*** Washers, lubricators, furnaces, and other conditioning equipment are considered a part of the actual fabrication process, and they are essential. In the case of the small job shop, parts are frequently sent to a specialist facility for the required function. Heat treating, plating, and painting are outside operations for the most part. Large companies that make parts on a continuous basis are likely to have their own facilities, as are companies that can make a number of parts on a batch basis that are similar and then run them through after enough have been accumulated. Storing parts to be batched tends to run counter to the idea of just-in-time, and the need for flexible designs that can be computer controlled is starting to appear in the conditioning equipment field. Where cycle time, sequencing, or heat history determines the configuration of the equipment and batch sizes, it may not be a simple matter to satisfy both requirements of just-in-time and adequate economic throughput. It is not an easy challenge.

6.9 Summary

Material handling is an important part of any CIM operation. The main function is to get the tools and workpieces to and from the machine tool. Excessive material handling is costly and undesirable, but some material handling is inevitable. Because in a CIM plant we wish to keep WIP to a minimum, effective material handling needs to be computer controlled.

There are many kinds of material handling equipment, including the loader or unloader, which is frequently a robot. Transporters such as conveyors, accumulators, and AGVs make up a large segment of material-handling equipment. Computer-controlled AS/AR systems are being used in many large factories.

Certain pieces of ancillary equipment cannot be conveniently programmed for small lot sizes and so are difficult to incorporate into a CIM facility. Ovens, washers, and plating equipment fit this category, as these are all continuous-process machines. Scheduling so as to put similar-sized products through these processes will help.

6.10 Exercises

1. What is the difference between a machine tool, a tool, and a tool holder?
2. What is included in a setup?

3. How is the choice of conveyance dependent on lot size?
4. What do loaders and unloaders do?
5. How are robots used for loading and unloading?
6. What do positioners do?
7. How is automatic gauging used on an assembly line?
8. Why are programmable controllers sometimes used in place of relays?
9. What is the purpose of an accumulator?
10. How does the blue-steel accumulator differ from a magazine or clip accumulator?
11. Why are robots used in palletizing?
12. Name three or four different kinds of conveyors.
13. Why don't we just eliminate storage?
14. What does AS/AR mean?
15. What does picking mean?
16. What do we call it when we put the subassemblies into packages for further assembly?
17. How are large parts in a warehouse usually stored?
18. Describe a bin operation.
19. What is one thing that vendors can do to help out in terms of a production process?
20. What is the best thing to do to ensure that things work right in the warehouse area?
21. How are very small parts used in high-speed assembly frequently packaged?
22. How does an AGV work?
23. What are the advantages and disadvantages of reflective tape and embedded wires for AGVs?
24. Why is the AGV aisle always clear?
25. What kind of equipment is usually relegated as an afterthought in plant design?
26. What are some of the problems in fitting washers, lubricators, and platers into the factory flow?

7 CIM Units: Computer-Aided Functions

7.1 Introduction

Computers and Humans

We have looked at computer-driven assembly equipment, automation and transfer equipment, and a specialized machine tool, the robot. Other programmable devices, whether micro, mini, or maxi, are important to the successful operation of a CIM facility, so we now turn our attention to these devices. All can be grouped into the class of computer-aided "something." Thus any machine that serves any function and that is a programmable device fits this mold. The major classes are computer-aided engineering (CAE), computer-aided design (CAD), computer-aided testing (CAT), and computer-aided manufacturing (CAM). Purists would stop with this list, although they might want to add a subset, computer-aided drafting, to the design function. Included are not only specific machine-based disciplines but also systems functions as well. These include but are not limited to scheduling, purchasing, inventory control, loading, and the like. In fact, anything should be included for which a computer could be used to help humans do their job more effectively. Humans need to be able to understand the process, model the process mathematically, and articulate the process. The computer does a good job of executing the process model, given the proper inputs. Most problems with computers can be traced to a fundamental lack of understanding of the process, the inability to properly model the process, or poor data. As Pogo said, "We have met the enemy and he is us!"

7.2 Computer-Aided Manufacturing (CAM)

▶ ***The Customer Determines What Is Made.*** The end result of activity in a manufacturing plant is the fabrication and assembly of a something. Any number of things go into the making of this something, but the fact remains that at any given time only one thing is produced, whether a molecule of matter out the end of a pipe or a car made of metal. The last stage is putting this thing in a package of some sort ready for shipment. In each stage of the process there are many inputs, but only one output. It is for this reason that these units of a CIM operation will be examined from back to front. What is manufactured is the focal point of all the activity that precedes the manufacturing. It makes economic sense to look at the situation from the viewpoint of what the customer wants and figure out how to get there, rather than the other way around. Furthermore, at this time anyway, most companies that are looking at CIM are in the material cutting and forming industries, such as automobiles and appliances. Most of these companies are involved with metals of some sort, but the principles are also applicable to such things as plastics and ceramics. Also, CIM concepts should not be limited to manufacturing, for there is a good argument for using CIM technologies in mining, on the one hand, and the service industries, on the other. To keep things simple, manufacturing in this context means modest-sized batches of a diversified product mix—*modern* cars and freezers, for example.

Numerical Control

A lathe is a machine tool used for turning round objects. The part of the lathe that holds the metal or workpiece is the chuck. Mounted on the lathe is a tool holder, which holds something that actually cuts the metal. This cutter is called the tool. Other machine tools include milling machines, drill presses, and grinders. Other tools include mill cutters, drills or bits, and wheels. On a lathe, the workpiece is rotated by the machine, and the tool is moved in until it touches the workpiece. Chips of metal are removed as the tool holder continues to move along its prescribed path. Coolant is used to remove the heat and to lubricate the tool end where it contacts the workpiece. There is a set of conditions where just the correct amount of material is removed by the tool during each pass. If not enough metal is removed, production will drop due to inefficiency. Likewise, if too much metal is removed, production will also drop because the tool will wear excessively and there will be waste. The production rate is dependent on the speed of the workpiece, the feed rate of the tool, the metallurgy of the workpiece and the tool, the kind and amount of the cutting oil, and many other factors.

Early wood lathes rotated the workpiece and the operator moved the chisel or tool in to the workpiece by hand, resting the tool on a tool rest. The rest was later replaced by a holder for metal cutting, and this was in turn moved in to the workpiece by a lead screw on dovetail ways, one of the truly great modern inventions. The next step was to motorize the tool holder for uniform feed rates and to cut threads.

Because, for any particular job, there is usually a set of specific conditions and sequences, it is possible to establish a procedure for the process and to accurately designate positional coordinates. With numerical control (NC), we use the positional coordinates to instruct the various parts of the machine tool where to go for a particular job. If we use a computer to effect control of a machine, it is called CNC. Most CNC machines use a perforated paper, plastic, or metal tape to provide the programming instructions to the programmable device mounted in the machine tool. This tape is usually prepared off line by engineering or production management. If a central computer is used to provide programming instructions to the machine processor, it is called direct numerical control (DNC). Figure 7.1 shows a typical NC process in simplified form, and Figure 7.2 shows the application of the tape to the machine.

Hierarchy of Control

A popular misconception is that in a computerized factory there is one huge computer that controls all the machine functions on all the machines. Early experiments along this line demonstrated conclusively that the concept does not work. What does work is a distributed system. In each machine tool there may be several levels of processors arranged in a hierarchial control system. At the lowest level would be functions such as controlling the mechanisms and drives that position the cutter head and the speed of the chuck. The next level might employ feedback sensors to ensure that the tool actually arrived at its destination and to direct the positioning of the cutter. The third level would delineate the tool path for a given job such as contour cutting. At each level the how-to-do-it is controlled by the processor at that level, but the what-to-do is controlled by the processor at the next higher level. Figure 7.3 shows the National Bureau of Standards hierarchial model going all the way up to a total CIM facility. A computer or programmable device on the third level would not be concerned with the routine functioning of the device on the first level unless there was a problem situation like a blown fuse or tool failure, in which case an emergency override procedure would be automatically instituted. Of course, a human programmer must have built this feature into the system.

As is true of everything having to do with computers, each manufacturer of CNC machine tools employs a different dialect for

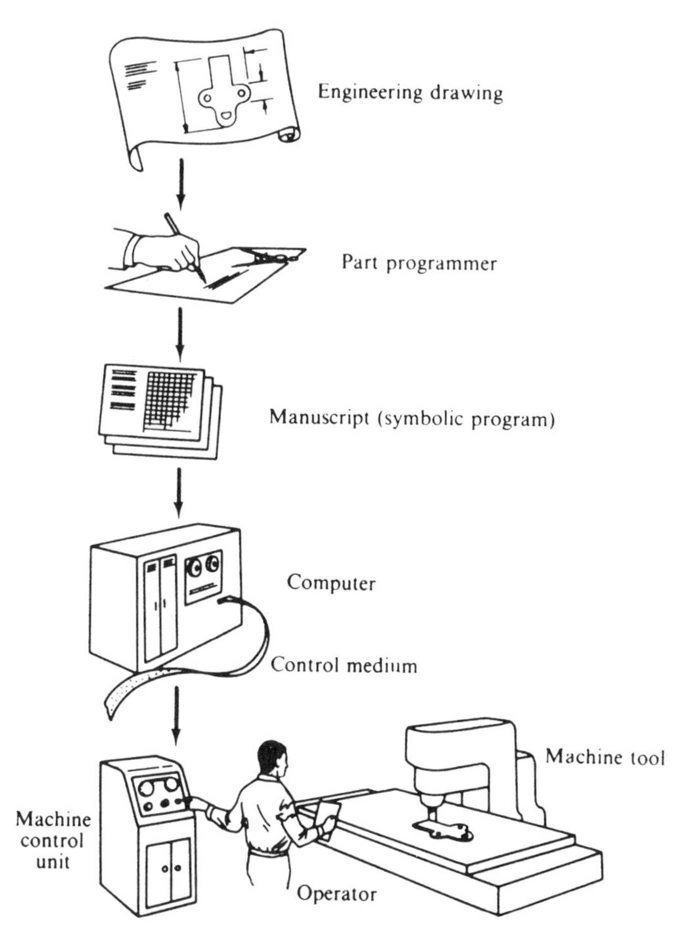

▶ ***Figure 7.1*** A Simplified Explanation of the NC Process (Bateson, *Introduction to Control System Technology*)

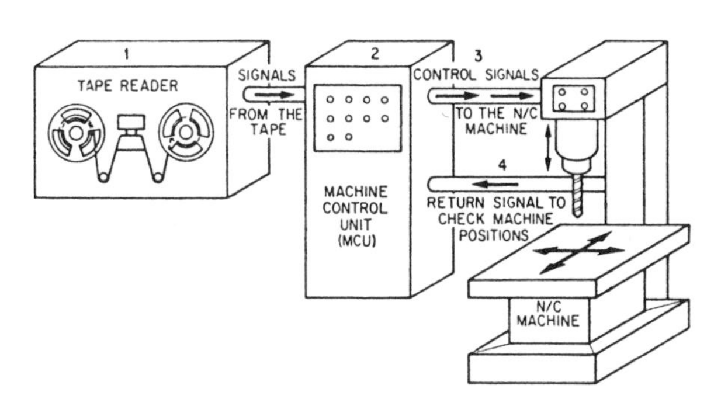

▶ ***Figure 7.2*** The Process of Converting Tape Codes into Machine Action (Roberts and Prentice, *Programming for Numerical Control Machines*)

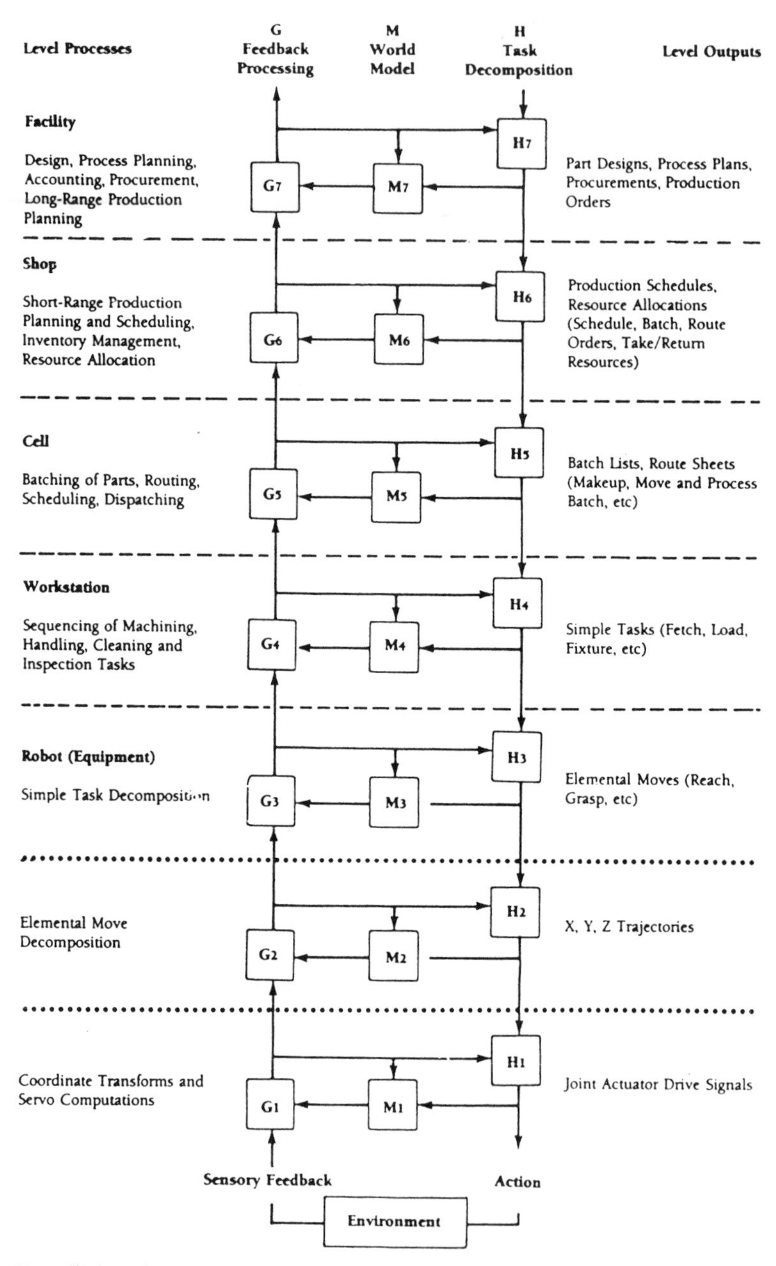

▶ ***Figure 7.3*** National Bureau of Standards Control Hierarchy (*CIM Review*)

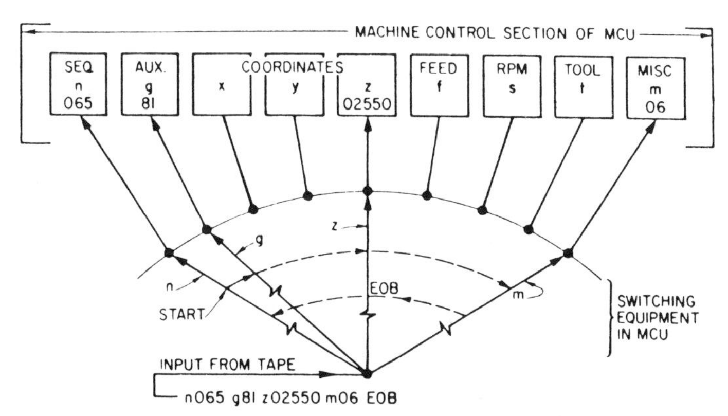

Schematic diagram of WORD ADDRESS programming illustrates the switching action called for by each letter in the block.

► ***Figure 7.4*** Word Address NC Programming (Roberts and Prentice, *Programming for Numerical Control Machines*)

language, although the situation is improving due to standards such as ASCII. Many are based on FORTRAN, with APT (Automatically Programmed Tool) probably being used more often than not. Generally, it is possible to specify information in XYZ coordinates for where the tool is to start and where it is to end and to incorporate information regarding rate of travel. Special subroutines are provided that will direct the tool to follow arcs of different-sized radii. Other instructions tell when to turn the coolant on, when to lift the tool from the workpiece, when to go to home position, and so on. In many cases it is up to the programmer to determine the width and depth of cut on each pass, feed rates, and other operational parameters. The programming of the machine tool requires skill and meticulous attention to detail. This is why programmers are highly paid. Provision for incorporating automatic tool changing and gauging is also available. A schematic diagram of the process is shown in Figure 7.4. This one line of code uses word address to direct the tool to go to a specific Z coordinate.

What has been described is really a system using a programmable controller or microprocessor, which may or may not have been designed by the machine tool builder. If not, the controller is made by one of the major builders and sold as an OEM (original equipment manufacturer) part. The programmable controller is also used in the machine tool for sequencing.

Need for Flexibility

We can combine manufacturing operations in many different ways to achieve greater throughput and better quality. Keep in mind, if we can make very large numbers of the same part or assembly, we will probably opt for hard automation. That is, we will use machines in which the machine, the tools, and all the infeed and outfeed conveyors are dedicated to making one part or assembly. These part-specific processes require a very large capital investment, but they will operate for many years with only minor modifications for model changes. If, due to competition or other reasons, we are compelled to change models frequently or to offer many different models in the same year, then an entirely different strategy is required. We need equipment that has inherent flexibility so that we can make changes quickly. Not only do we need to incorporate a variety of model changes in our design, but we need to be able to make each model in comparatively smaller batch sizes and with very rapid process and tooling changes for each model. We do not want to get stuck with 500,000 parts that we cannot sell, nor do we want to miss out on a market because we could not tool up fast enough.

Manufacturing Cell

There are two popular approaches to making a variety of parts quickly. The first is to design a manufacturing process around a manufacturing cell. This is also called the product type of layout. Although it has been used for quite some time, it is not as widely used as the colony or process layout. This more traditional approach in a manufacturing operation centers all similar machine tools in the same area. All the lathes are in one room, all the millers in the next, and so on. The problem with the colony layout is that, if a part requires a series of operations, the part must be transported to a variety of machine stations and wait its turn for processing at each station. It is not uncommon under this procedure for a part to spend 2% of its time being machined, 3% of its time for tool change while in the machine tool, and 95% of its time waiting in a queue to get into the machine tool. Something that takes about a week of actual machining and setup time could end up taking three months to work its way through the process.

A manufacturing cell, on the other hand, does all the operations on the part at the same location. The robot is an invaluable aid here because it can transport the part easily and quickly from one machine tool to another. Also, because different parts use a different sequence of operations, the robot can be programmed to deliver the part to the right machine at the right time. One sequence might be turn, mill, and drill, and the next sequence might be drill, mill, and grind. If the robot is in the center of the cell and the machine tools are around it in a circle, the sequence could be in any order. Thus the cell approach eliminates the waiting time between machine operations. For it to be effective, we still need to run the parts through in reasonable-sized batches, because it is still necessary to change tools and set up for different parts. However,

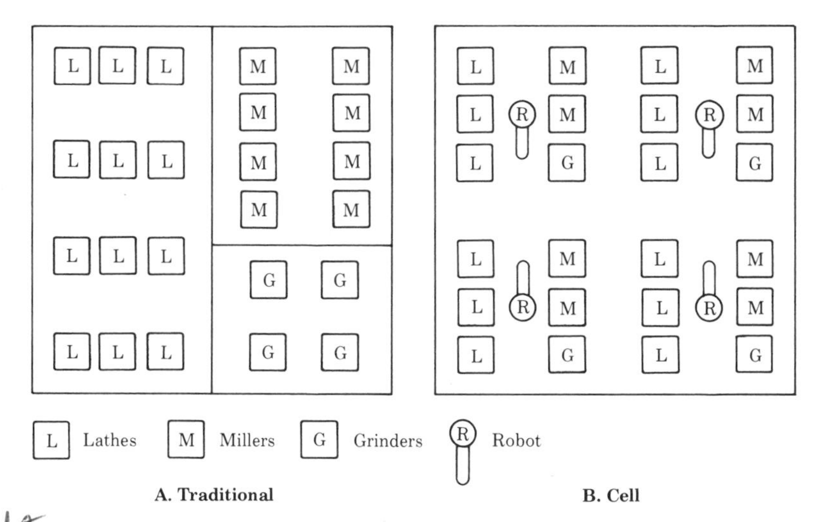

▶ ***Figure 7.5*** Traditional versus Cell Approach to Manufacturing

we are able to accommodate a fairly wide variety of parts. A model of the traditional colony versus the cell approach to manufacturing is shown in Figure 7.5.

Flexible Manufacturing Systems

Another approach to the problem is through a flexible manufacturing system (FMS). With FMS we use dedicated pieces of machines, again in a cell, but we make all sizes of parts in the same class. We make the parts in random fashion. To accomplish this, we need to design the parts using what is known as group technology. Because of product evolution and for political reasons, most products are designed from start to finish without much thought being given to whether similar parts in different models or even completely different appliance parts can be made easily on the same machine. The part is designed and then the process is determined. From the process, the machine tools are selected to make only that part. However, if the design functions in all appliances were coordinated, then all the shafts, for example, could be designed with similar enough features so that one machine could make all the shafts. A similar shaft could be used in the vacuum cleaner, mixer, and dishwasher, or the Chevy and the Pontiac. A FMS facility could then be programmed to make two vacuum cleaner shafts, six washer shafts, three mixer shafts, and so on, in random fashion, depending on what was needed at that particular instant. Although group technology begins with the initial design stage, the philosophy of group technology needs to be used in all phases of manufacturing. This will help to ensure that all products that can fit into the group actually end up in the group.

Workings of an FMS

Figure 7.6 shows an FMS setup offered the W. A. Whitney Company. Material, in this case sheets, is off loaded onto the FMS bed, where the part is blanked, transferred to a holding area, and then formed into the finished parts and placed into storage. The parts, although different, are similar enough to be made on the same machine. Note the rack for the tooling dies. Not shown is the computer that runs the system. For FMS to work, several things have to happen in addition to correct product design up front. Different-sized workpieces, different workpiece holders, different tools and tool holders, and a different program of operations need to arrive at the machine tool at the same time to make the part. After the part is made, the finished part, tools, and holders must be removed so that the next part can be made.

For the next part, raw stock needs to be conveyed to the machine tool, and the correct-sized holder has to be brought in from a crib and mounted on the machine tool. Concurrently, the tool holder with tool must be brought to the machine and positioned correctly. The machine tool must be programmed correctly to make the part. Automatic gauging and tool sensors must be located to ensure that the part is being fabricated correctly. We need to change tools frequently and quickly while we are making the part because there may be several operations for a given part. We need conveyances for the tools and holders, as well

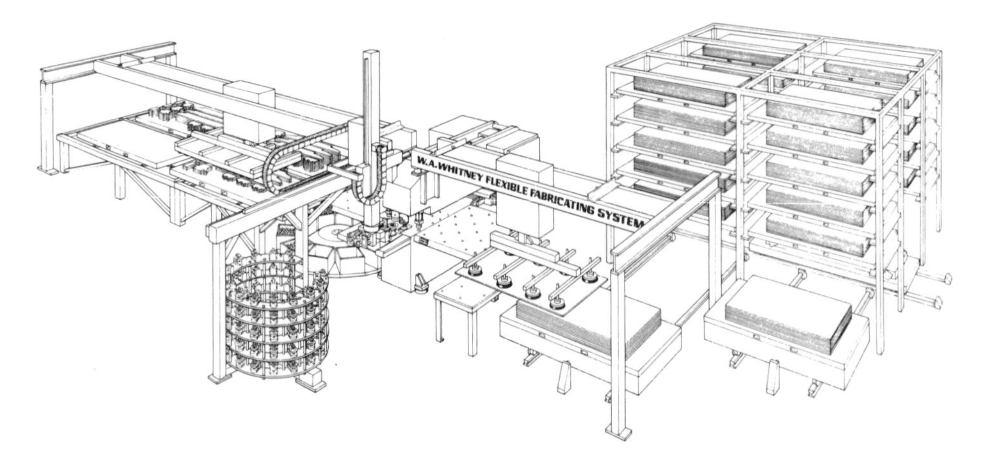

► ***Figure 7.6*** A Flexible Manufacturing System (Courtesy of W. A. Whitney Corp.)

as for the raw stock. Usually, tools are mounted in a tool holder off line and placed into a storage bin until needed.

As can be seen, an FMS needs a very large programmable device to run it and requires a very large investment; $10 million is not unreasonable. The FMS needs enough work to keep it busy all the time and must operate without any downtime (100% reliability) to be economically feasible. Perfect planning and scheduling are required, and this is usually done by a higher-level computer. Because of the practical inability to meet these very severe requirements, FMS has not achieved either the market or the growth rate anticipated earlier by industry forecasters.

7.3 Computer-Aided Testing (CAT)

At the production end of the engineering–production continuum, we find testing. Testing ranges all the way from simple go/no-go gauging to testing under dynamic conditions and simulation. Testing of parts can be on or off line. When on line, testing is usually nondestructive and uses vision systems and/or contact and noncontact gauges. Off-line testing can be both nondestructive and destructive. Sampling and statistical techniques are usually employed. In modern plants the operator is frequently trained in these techniques.

► ***Feedback Is Used in CAT.*** Real-time testing or gauging is used to provide feedback control information to the manufacturing process. In a typical setup, a sensor such as a LVDT is mounted on the tool holder, and dimensional measurements are made on the workpiece as it is being operated on. If too much metal is removed, or feed rates are exceeded, or any other criterion is violated, an error signal is generated, which results in a correction being incorporated into the process. This type of control is handled by the local programmable device and is used to ensure the output is within acceptable limits. When enough metal has been removed, the controller turns off the machine and calls for the next sequence in the cycle. The controller can also be programmed to do something like sound an alarm if a tool breaks or too much metal is removed.

Many times it is necessary to measure the workpiece under dynamic conditions. In a gas turbine, for example, the blades deflect while running due to dynamic forces. Tolerance specifications are established for the amount of allowable deflection at a given rpm and under certain other operating conditions. Built-up subassemblies are placed on very precise dynamic dimensional gauges, and the turbine is run according to a prescribed routing. The test stand in this instance is completely computer controlled, and noncontact capacitance sensors are employed. In a similar sense, LSI (large scale integrated) circuit chips are cycle tested in very complex machinery to ensure they do what they are supposed to do under various duty cycles and in extreme environments. The design, fabrication, and marketing of automatic test equipment (ATE) is a large and diverse industry.

▶ ***Close the Control Loop.*** Any gauging, whether manual or automatic, computer controlled or not, is designed to provide information. The information needs to serve a valid product-acceptance function, and many gauges are employed to tell the operator or management that a part meets a certain specification. In this case, management will buy the gauge only when their process is out of control or has the potential to be out of control. However, the real savings in producing quality parts comes when gauging is used to control the fabrication process itself. Not only does the gauge measure the parts, but by providing an error signal the process can be adjusted to make consistently better parts.

Valve springs, such as used in automobile engines, are made by impinging a wire against a tool called a point. The point deflects the wire and the wire comes off the point in a helix as a freestanding coil. At the right time, a cutoff tool cuts the spring off the wire and the spring drops into a bin. A noncontact gauge, made by a very small manufacturer, was developed to measure the free length of the spring after coiling but before cutoff. Real success came to this company when the gauge was then used to sort good springs from bad and to adjust the cams so that the next spring came out right. Now the customer produces a very large box of acceptable springs that has been 100% inspected and sorted and a tiny box of rejects. Because the gauge system is so precise, it is now impossible to make these springs without this gauge. Whereas before the gauge was difficult to sell, when it was transformed into a controller gauge, production could not keep up and the customer was willing to pay almost any price. This kind of gauge is known as a coordinate measuring machine (CMM). CMMs are used widely in the automotive industry. They measure and control dimensional characteristics such as diameter, concentricity, and runout. The power in automatic gauging lies in closing the control loop.

▶ ***Readout Can Be Varied.*** In large test setups, several hundred individual pieces of data may be generated and read simultaneously. The signals from the sensors are usually analog. To be useful to a computer, the signals need to be conditioned with signal conditioners and read with data loggers. Under the dynamic conditions of velocity and acceleration measurements, for example, additional circuitry, such as peak read and sample and hold, is provided. These signals also must be converted into a digital format. Sometimes data conditioners and loggers read the data points in a sequential fashion, but more advanced units can be programmed. The information can be presented in absolute terms through a display to the operator as exceeding preestablished upper or lower set

▶ ***Photo 7.1*** Automatic Testing Machine (Courtesy of Texas Instruments Incorporated)

points, or as analog or digital output for interrogation by a higher-level machine. Historical data can thus be further processed statistically for trend analysis, parts count, productivity indexes, and many other things, depending on what management needs. Photo 7.1 shows a computer-aided test setup for transmission design using a Texas Instruments computer.

7.4 Computer-Aided Design (CAD)

▶ ***Human Creativity Is Still Needed.*** On the other side of the factory, so to speak, we find data input functions going into the manufacturing operation. While the programmable devices on the machine tools will be extracting information from the data base on what and how many things to make, as well as specific instructions on how to make them, the input programmable devices will be loading the data base with the information the machine tools need to function. Computer-aided design assists us in the proper design of the components of the product, as well as in the final assembly. The two words "aided" and "assisted" are important, because they say that the creation of the design concepts needs to be human. We still need smart people to figure out what we are going to make and how we are going to make it. CAD removes the tedium in drafting, ensures mating parts will fit together, keeps track of hundreds of details in the data base, and generally speeds the process of generating working drawings. Another bonus of computerized drafting is that it simplifies making drawings for similar products and facilitates group technology.

From a practical viewpoint, CAD replaces the detail draftspeople and allows the designer to go directly from idea to print using automation. A large CAD system is capable of replacing a number of draftspeople, and even though a large system is costly the payback period is usually attractive. It is for this reason that the CAD/CAM element in a CIM environment is almost always the first unit to be put in place. As a result, there are a number of large manufacturers of the necessary equipment and the technology is fairly well understood. However, installing a CAD/CAM system is still very time consuming and difficult. A clear statement of objectives and a well thought out implementation program are essential. General Electric Aircraft Engine group has been working on the problem for over 5 years and is not yet finished.

► ***Different CADs Use Different Computers.*** The heart of a CAD system is the computer. Depending on system size, the computers used range from fairly extensive mini computers, such as Digital's Vax system, down to micros like the IBM AT or PC. A typical work station from Control Data is shown in Photo 7.2. Very large mainframes are not used too frequently. The reason is that, although large memories are required, it is not usually necessary to access the memory rapidly, such

► ***Photo 7.2*** Typical Work Station (Courtesy of Control Data Corporation)

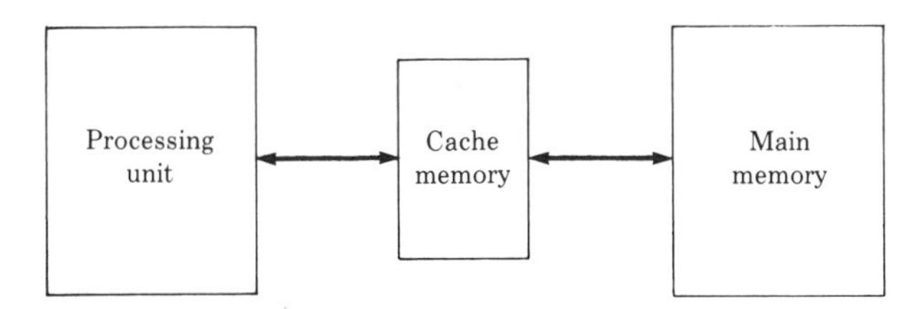

▶ ***Figure 7.7*** Cache Memory in a CAD System's Processor (Goetsch, *Computer-Aided Drafting*)

as might be required with pure number crunching. There are ways to access external memory, such as a disk drive, and fool the operator into thinking the entire program is in the machine. This concept is called virtual memory. Segments of the program are brought into memory and overlaid on what is already there as required. Another idea is to use processor chips that support a cache memory. A cache is a small memory with rapid access time that stores the frequently used routines. Figure 7.7 shows this idea. Seldom-used routines and data are stored in the main memory of the computer and can be transferred either to the cache or directly to the processor, depending on circumstances. Data such as completed drawings are generally stored off line on tape or other media to be brought into on-line storage or memory when needed. One trend particularly worth noting is that, as microprocessors become more powerful and cheap and disk storage becomes larger, more and more systems are using inexpensive personal-type computers. In addition, most of the hardware and software in the PC computer supports compatible operating systems and can be uploaded and downloaded easily for access to larger data banks. It is estimated that about 70% of all CAD requirements can be handled by microprocessor systems that cost less than $20,000.

▶ ***We Need Data and Graphic Entry Systems.*** The computer hardware can be located anywhere, but the place where the operator sits is the work station. Here, we need interface equipment to get meaningful data into the computer. A variety of specific pieces of hardware are useful and many options are offered. The basic method of numerical data entry is either a full keyboard or a keypad. The keyboard looks like a typewriter keyboard, whereas most keypads look like an adding machine keypad. Both are used for entering text and numerical information. Frequently, a monochrome monitor is employed that is especially designed for character readout. Keyboards are not particularly suited for entering drawing information, and a monochrome monitor is not suited for the display of graphics.

A digitizer is a device that lets the designer move a hand-held device along a flat surface as you would when you are drawing something. Common devices are pens, a trackball (mouse), and a puck. A puck has cross hairs and buttons for command purposes. Most digitizers work on the principle of embedding many wires into the table in gridlike fashion and using the puck to generate a magnetic field over the wires. The location of the field is what is fed into the computer. A dot of light displayed on the screen shows where the puck is at a particular time. The designer pushes a button on the puck to record that position. Some designers prefer to use a pen rather than a puck.

► ***Most CAD Systems Are Menu Driven.*** Also located at the work station, and frequently on the digitizer, is a graphics menu that allows the designer to quickly generate internal functions in the computer, hence the name functions menu. Figure 7.8 shows a typical functions menu for isometric dimensional drawing. Pictures and symbols of each function available are shown on the menu, usually each in its own little box. By pointing the pen to a particular function and pushing the button, the designer can generate the function without having to draw it. For example, the designer locates the center of a circle on the drawing, points to a circle on the menu, and draws the circle by pushing the button. The circle can be enlarged or reduced, dimensioned, shaded, or what have you, simply by pointing the pen and pushing the button. Different function menus are available, such as for making isometric drawings, for adding text and dimensions, and for industry-specific purposes such as circuit design or piping layouts. Photo 7.3 shows a typical CAD setup. Note the digitizer, pen, and display.

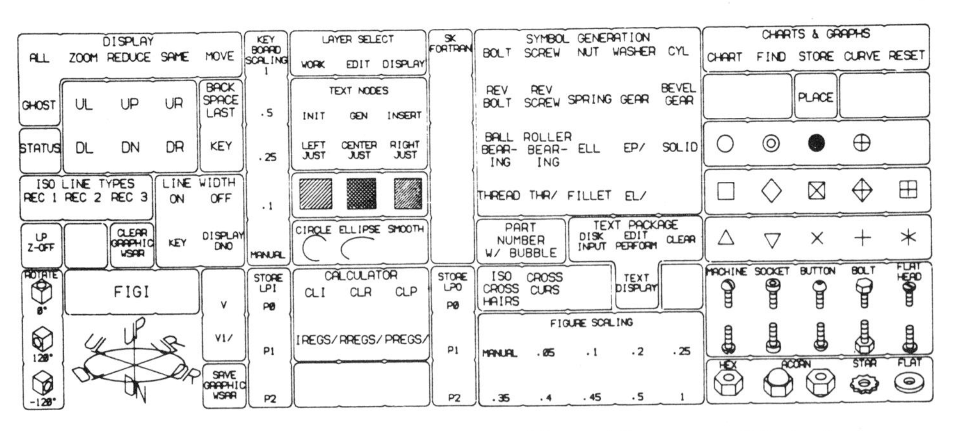

► ***Figure 7.8*** Isometric Drawing Menu (Courtesy of Auto-trol Technology Corporation)

▶ ***Photo 7.3*** Typical CAD Layout (Courtesy of John Deere & Company)

Types of Displays

The CAD system is an interactive system. This means that it is necessary to provide the designer or operator with an instant visual display of what he or she is doing. This display, or soft copy, is perhaps the key element of a successful CAD system. The typical character display mentioned earlier is similar to a TV set in which a beam sweeps across the screen both horizontally and vertically and generates a raster, or series of dots. Lines are generated by turning on contiguous dots. Each dot is called a pixel. The closer each dot is to its neighbor, the higher the resolution. This is sometimes called a raster graphics display. To get clear, accurate lines, we need to use high resolution, say 660 by 880 pixels. But for many applications, this is not good enough.

Another type of display is a directed beam-refresh graphics display. Here we direct the beam to a continuous phosphor coating that also glows. The glow life of the phosphor is very short and so the line needs to be continuously redrawn very rapidly. To get around this, a storage tube was developed that contains a backplate on which the

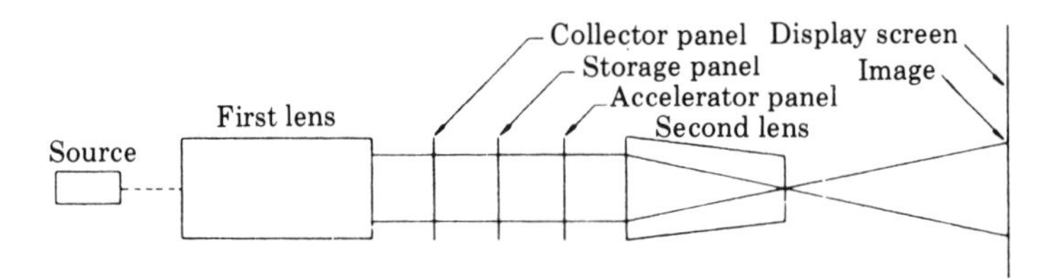

A. Direct beam-refresh display composition

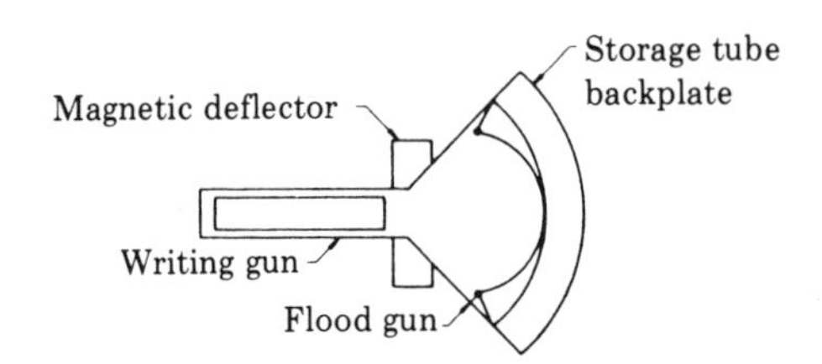

B. Direct-view storage tube display composition

▶ ***Figure 7.9*** Graphic Display Systems (Goetsch, *Computer-Aided Drafting*)

image is drawn and stored. In this way, each individual line does not need to be continuously refreshed. Advanced models allow you to add and subtract lines from the storage tube incrementally, so the display does not have to be refreshed or redrawn each time something is changed. All displays are available in color models as well as monochrome. The more features, the higher the price is. Two types of common displays are shown in Figure 7.9.

▶ ***Printers Give Us a Hard Copy.*** There comes a time when we need to have a hard copy or permanent print of what we have generated. Two general classes of devices give us a hard-copy output. For quick and simple things, an electrostatic printer, such as an ordinary office copier, is used. These copies are used to check ideas and to allow the designer to plan changes. They are not intended as high-quality working drawings or prints. For this, we need to use a plotter. A pen plotter is essentially an XY robot that has a pen or pencil attached to the gripper. Plotters are available in various sizes and can draw with several different colors of ink at the same time. Another type of plotter is a photoplotter that uses a light beam instead of a pen and draws on sensitive paper. The mechanisms are similar and use either an XY table or the Y is replaced by a motor-driven drum.

7.5 Computer-Aided Engineering (CAE)

It is very hard to differentiate the boundary between engineering and design, and it is likewise hard to differentiate the boundary between design and drafting. Rather, think of them as a continuum without any boundaries. Conceptually, certain tools are available to CAE that are not available to the CAD system. However, most CAD systems have CAE capabilities as well.

► ***Programs Are Available to Do Engineering.*** Since most work stations have a keyboard entry terminal, display, and a conventional printer, it is a simple step to make available the necessary programming so that the engineer can do scratch pad or even more advanced calculations. Here the key word is program. Since all the hardware is usually in place, the question becomes one of what programs are available for the engineer's use. For example, a series of programs, specifically graphics oriented, is finite modeling or finite element analysis. This allows the engineer to draw a pictorial three-dimensional representation of the part using a wire-frame drawing. A wire-frame drawing outlines the part with a series of topographical lines like those found on a contour map. The intersections of the lines are made to represent some physical characteristic, such as stress. Deformation of the part is simulated in the computer and the resultant drawing shows the design engineer what locations on the part are liable to result in failure. Sometimes stress and other characteristics like temperature are displayed as areas of different colors. New programs are being offered all the time. In the aerospace program we sometimes need to put small containers inside of a larger container. This is called equipment packaging. If you have ever taken a trip with your family, you know what it is like trying to get all the luggage into a too small trunk. Computer modeling and simulation are useful tools for this type of activity as well.

7.6 Human Factors in CAE/CAD/CAM

The whole idea of a CIM plant is to increase productive output without sacrificing quality. The more automated equipment we include the greater is the need for considering the human factors and ergonomic characteristics of the interfaces between people and machines. One area where a great deal of attention has been focused recently is video display terminals and all kinds of data entry. More enlightened companies are interested in having well-designed equipment to enhance productivity. Unhappily, most of the impetus for well-designed equipment has come from the unions for health purposes. Back pains and eye strain have been the two most common complaints. In any event, the result is that users of equipment have been directly involved in the specification and design of interface equipment supplied by the various vendors.

Human factors have been extensively studied in CAD but not CAM. If you walk into a CAD room, you are likely to find people sitting in comfortable chairs that are completely adjustable for their particular body. However, the first thing you will notice is that the room is quite dark, because studies have found that for bright lines on a dark screen this type of lighting is best. Because different screens and drawings have different light outputs and because each individual has preferences, there is usually additional provision for adjusting the light level locally. The layout of the keyboard, digitizer table, and other pieces of equipment is again based on extensive studies. Monitor screens can be raised, lowered, rotated, and tilted, within limits, and the picture can be adjusted for brightness, contrast, and other things, just like a TV set. Sometimes it is desirable to have black lines on a light background, so provision is made to toggle this feature. Usually, the curser blinks to show where it is. The blinking rate is important, as well as the refresh rate of the images.

Out on the factory floor there is a different situation. At present, not too much attention has been given to the ergonometrics for the factory worker except in critical cases like a gauging or testing area. We still see operators sitting on hard seats or no-back stools while tending machines. This may or may not be OK, but who really knows? The point is that this area has not been studied as extensively as the situation in the CAD room.

7.7 Inventory Control and Scheduling

For CIM to be effective, we need to adapt a just-in-time (JIT) philosophy. It would be possible to incorporate automation or CIM without JIT, but it would be very difficult. This is because CIM ties all departments together. Tandem operations without JIT result in a facility that operates with islands of automation. It should be stressed that JIT has to come first and then CIM; otherwise, it is money lost. For just-in-time to work, we need to have control over inventory and scheduling. The more perfect the control, the better off we are. Of all the variables that enter into good inventory control and scheduling, perhaps data entry is the most difficult to resolve. This is mainly because people still need to enter most of the data and people make mistakes.

With the exception of bar-code readers, nearly all data are entered manually through terminals. For production scheduling, a vital piece of information is the production order, which results from sales to customers. Under just-in-case operation, we made a lot of finished goods and held them for the customer's order. If something went wrong, we could probably get away with shipping a like good out of stock without penalizing the customer too much. Under just-in-time shipping, the wrong merchandise means we have to go through the entire cycle before the customer gets what he or she ordered. Everything should be correct on the order when it arrives and it needs to be entered properly. We can use redundant data entry and other techniques to minimize the problems, but at considerable expense. There is great promise for voice data entry to speed up the process, but we have a long way to go.

Under the ideal JIT scheme, the material on order comes into the shop concurrently with the sales or production order. This means that purchasing must enter the orders with the vendors correctly. One way to reduce the risk of transcribing errors is to tie the two computers together. In terms of the physical movement of goods, many companies are asking their vendors to affix coding labels for machine reading. However, in most cases the goods are manually labeled at the receiving point before being put into inventory.

7.8 Summary

The implementation of CIM is an evolutionary process, so some factories will become more integrated before others. However, all modern factories will use an extensive number of computers to aid individuals in various functions. As long as the computers are only being used as a passive information source, the system is not integrated. When we use the output from one computer to effect the operation of another computer, the system becomes integrated.

Electronic digital feedback is an essential part of the integrated operation. CMM machines are used to control the machine tool and other processes, and CAT machines operate on line to provide 100% inspection and control.

On the factory floor we find machine tools either directly controlled by a computer or a series of machines controlled by a computer hierarchy. This is called CAM. CAD systems assist the design engineer in developing the parts to be subsequently fabricated. When we tie the CAD system into the CAM system, we have integrated the two functions into a CIM element called CAD/CAM.

7.9 Exercises

1. Define CAD, CAE, CAT, and CAM.
2. In a metal-cutting operation, what is the optimum condition?
3. What is the difference between NC, CNC, and DNC?
4. What do we mean by hierarchy of control?
5. What modern conditions force us into making smaller batch sizes?

6. What is a manufacturing cell?
7. What is meant by group technology?
8. What four things need to happen for an FMS system to operate properly?
9. What is computer-aided testing supposed to do?
10. What is a coordinate measuring machine?
11. What does automatic test equipment do?
12. Why do we need to precondition data?
13. Can we eliminate human beings with CAD?
14. Which employee group may be eliminated by CAD?
15. What, generally, is the first piece of equipment to be put into a CIM environment?
16. What does virtual memory mean and what is cache memory?
17. What are the two means to enter data in graphics through the computer?
18. Name three common types of digitizing devices and describe them.
19. What is the difference between a raster graphics display and the directed beam-refresh graphics display?
20. Name two kinds of printers.
21. Name two typical types of engineering programs that are available.
22. Why are human factors important in CAD rooms?
23. Can you offer any reasons why ergonometrics has not been incorporated on the factory floor?
24. Why do we do inventory and scheduling control in a just-in-time environment?

8 System Design

8.1 Introduction

We have now described all the components that make up computer-aided islands, which many people refer to as islands of automation. Our robot is programmed to act as an infeed/outfeed mechanism for machine loading or unloading or it does a specific task such as welding. We use the computer to help us in design, scheduling, and the like. However, until we tie all these units together, we still have not accomplished the goal of the integrated factory.

8.2 Pareto's Law Applied

▶ ***Put Effort into Things That Are Worthwhile.*** There is nothing wrong with using a robot to do a simple task like loading and unloading and not taking any further steps toward integration of the factory if we are able to make cost-effective parts at this level. To go further, we must assume risks several magnitudes higher than at the level we are now. The payback must be commensurate with the risks. In fact, many Japanese plants stop at this level because they are able to use their workers effectively to do many of the things we are starting to relegate to the computer. There are several reasons for this, the main one being the cultural fact that there are strong interpersonal relationships between Japanese workers and their management. In any event, we must now take a close look to determine whether we will be putting the bulk of the available capital into an effort that will yield only marginal gains. We have to consider Pareto's law.

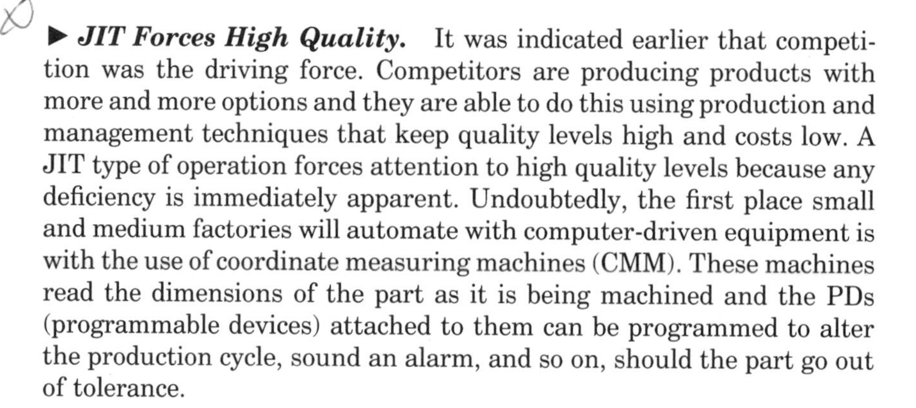

▶ ***JIT Forces High Quality.*** It was indicated earlier that competition was the driving force. Competitors are producing products with more and more options and they are able to do this using production and management techniques that keep quality levels high and costs low. A JIT type of operation forces attention to high quality levels because any deficiency is immediately apparent. Undoubtedly, the first place small and medium factories will automate with computer-driven equipment is with the use of coordinate measuring machines (CMM). These machines read the dimensions of the part as it is being machined and the PDs (programmable devices) attached to them can be programmed to alter the production cycle, sound an alarm, and so on, should the part go out of tolerance.

8.3 Cell Structure

▶ ***Cells Do All Operations on the Part.*** The simplest expansion of a single machine doing a single task is to have a cluster or group of machines doing a series of operations on a given part. We might have a lathe doing turning, a mill cutting, a drill making holes, and so on, and we would like to transfer the part from machine station to machine station. The robot is particularly well suited for this kind of transfer challenge. The raw part comes into the area on a conveyance of some kind, but once there the machines take over in the fabrication of the part according to the programmed process. Since we have a series of different machine tools in the same general area and since the operation is programmed, we can make parts with very little human intervention. This is called a cell structure. The cell structure concept is in contrast to the usual practice of having all similar machines grouped together and bringing all the work to the particular operation. Cell structure can be and is very efficient for making long runs of the same part. If a different part is wanted, either the tooling in the cell must be changed or, if volume warrants, a separate cell can be set up to make that part. In Figure 8.1 we see what the cell structure model looks like using a robot to handle the materials.

8.4 Group Technology

▶ ***Design Products with Common Parts.*** There is little communication between the product design group and the manufacturing group in many companies. Until very recently, manufacturing has taken a back seat to marketing, finance, and design in terms of getting to market with a product. However, this situation will change, and when it

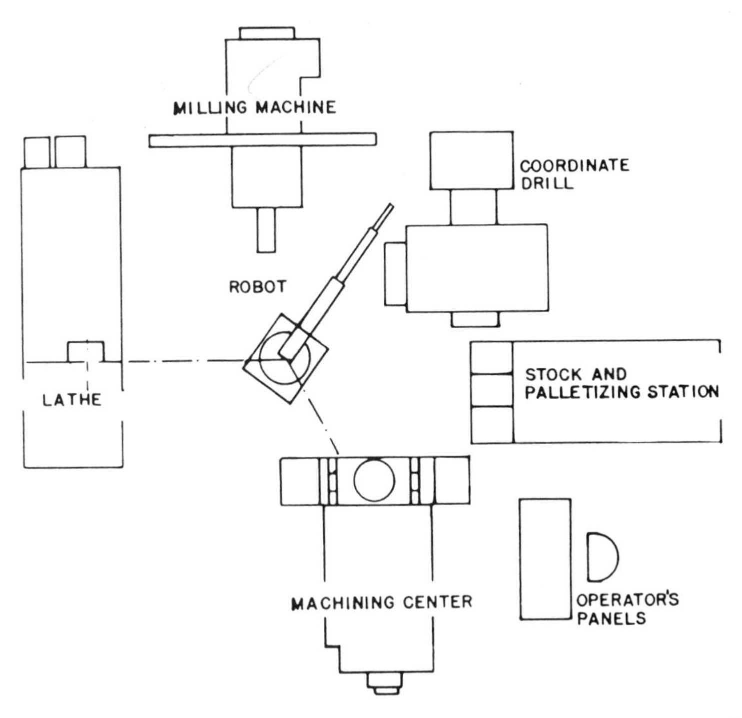

▶ ***Figure 8.1*** Technological "Core" of a Robot-Integrated Manufacturing Cell (Ayers and Miller, *Robotics*)

does the product design and manufacturing people will have to work together to come up with product lines that not only function in the marketplace but can also be made cheaply. The careful reader will notice the term product line. The customer wants products in different sizes and colors and with different options. Smart engineers will foresee the desirability of using similar components in all the various models of the same line. The individual parts will in all likelihood have different dimensions and features, but they should be like enough so that they can be fabricated on the same machines. We may offer the consumer 100 different disposer models, but we should be able to make the shaft for all models on the same equipment with only minor tooling and fixturing changes. We refer to the design of different parts to be fabricated on the same machines as group technology. A model of group technology is shown in Figure 8.2. In this model, we see that as the parts flow through the factory different machine centers handle different parts. Not all operations are performed on all parts. This is more realistic than to assume that one machine can handle all the functions of all the parts, unless it is a specially designed FMS system.

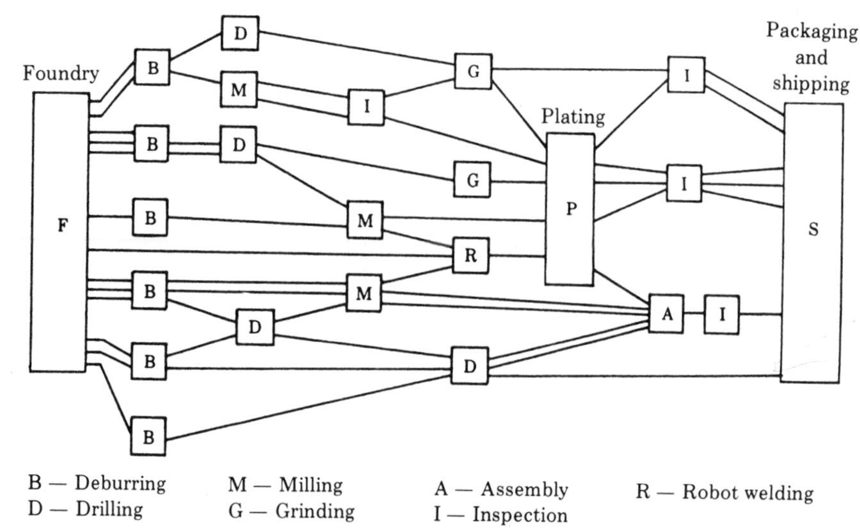

▶ ***Figure 8.2*** Group Technology Layout (Asfahl, *Robots and Manufacturing Automation*)

8.5 Flexible Manufacturing

▶ ***We Can Do All Operations on One Machine.*** Hand in hand with group technology and the JIT concept is the flexible manufacturing system (FMS). The JIT comes first, then the FMS. Otherwise, we would be wasting our investment in flexibility. This relationship will be discussed more deeply later. In the FMS concept we have a gigantic machine center that is completely computer driven. If production scheduling calls for one of a particular part to be fabricated, the machine will automatically set itself up to make one part, and the one part will be made. The next order from production scheduling may be for something completely different. Figure 8.3 shows the schematic of a FMS system. This line is for making various kinds of gears. Obviously, it is unlikely for the same machine center to make car engine blocks and door handles, but it could make small shafts with threads on one end and then switch over to large shafts with holes on the other end. In other words, the machine center must be capable of making a series of parts all in the same group.

▶ ***Make Stock and Tools Available at the FMS.*** The machine center must not only have all the machine tool components (lathe, mill, etc.), but it must also be able to move the raw stock, or workpiece, into the fixture or chuck, bring the correct tooling to the workpiece, and do the assigned tasks. When production batch sizes are large, the differences

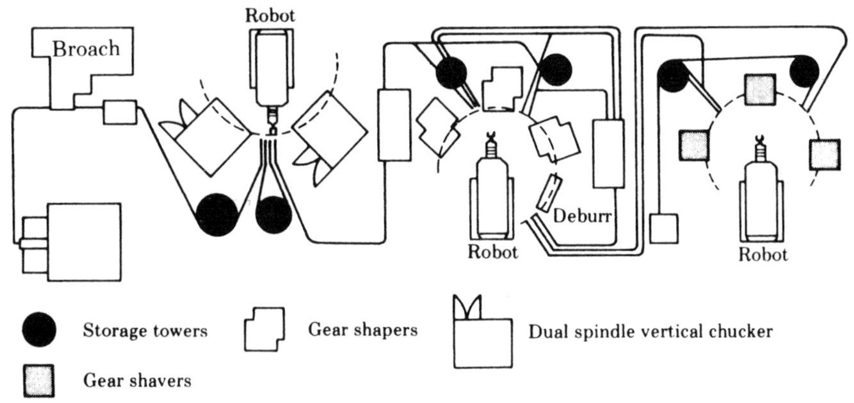

▶ ***Figure 8.3*** Flexible Computerized Manufacturing System (Ayers and Miller, *Robotics*)

between the FMS and a conventional automated transfer operation are not too apparent. However, when batch sizes become small and approach a batch size of 1, the FMS advantages become readily apparent.

Each product in the group has a complete series of specifications that it takes to fabricate the part. The process will have been set and the appropriate tooling determined. Each machine operation will have been broken down into its simplest elements. For example, if we are making a shaft, the raw material bar stock will have been set by the design engineers. The proper selection of the cutting tool will have been determined by an expert machinist, along with the necessary feed and coolant rates. The tool path, the locus of the center line of the tool in relation to the cutting edge, will have been worked out, probably on a CAD system. The tool holders and workpiece fixtures will have been selected. All these things plus many more details will have been worked out.

Once the process has been completely worked out, the programmer enters the process into the FMS computer so that operations may proceed sequentially according to plan. Once the button is pushed, the programmable device in the machine tool takes over and operation begins. The proper bar stock is brought over to the machine by conveyor or robotic mechanism and inserted in the holder. In the case of the lathe, the raw stock will probably enter through a hole in the center of the chuck, where it is clamped in place. The correct tool for the operation is loaded into a tool holder, if that has not already been done, and the tool holder is brought to the machine and mounted. The machine is turned on, the tool is brought to the workpiece, and chips are made as the tool cuts the metal. Coordinate measuring system gauges measure the work as it progresses. When enough metal has been removed, work stops and the machine sequences into the next part of the cycle. Each operation alone sounds simple. Tying them together has yet to be achieved commercially on a wide scale, although one or two showcase installations are operating.

► ***Part Selection Is Done by Computer.*** The FMS does not know which part will be fabricated next. Operationally, the cycle time for making the same part again as opposed to making a different part may be somewhat less, assuming there are the same number of steps, but not significantly so. The raw stock may not need to be completely removed and new stock brought in, and some other minor steps may be omitted. But, basically, the program and sequence need to be reset to start and the process cycle repeated. It takes the same amount of time for a tool holder that holds a 2-inch drill to get to the machine from the storage area as it does for a tool holder that holds a 2¼-inch drill. Actually, the tool holder with tool will already have been staged at the machine tool.

8.6 Hybrid Systems

► ***We Can Mix and Match Tools Depending on the Job.*** There are many ways to combine the various machines, technologies, and cell structures. However, the central theme is to use the computer-controlled machine where it does something best for that application and let humans do what they do best. Robots are very good at pick and place, part positioning, such as is found in component placement on printed circuit boards, and palletizing. However, they are not yet very well suited for general assembly work, such as screwing things together. For this reason, we find in machining centers such as the FMS that the FMS does the metal cutting, but the human selects and loads the individual tools into the tool holders and end effectors. The aforementioned drill needs to be inspected and checked; then it needs to be placed into the tool holder just so, and the drill is clamped with just the right amount of pressure. If the tool is very complicated, great care must be taken to do all the steps correctly. In other words, a lot of judgment is required in the process. The computer's lack of judgmental ability is why robots are not used in complicated assembly operations at the present time.

8.7 Integrating Vendors

Enlightened purchasing people are more interested in value than in price. And the days of playing one vendor against the other in terms of price only is rapidly disappearing, because large companies are finding that their vendors are simply an extension of their manufacturing operations. The flow of information to segments of the manufacturing area within the corporation is no more vital than the flow of information to the segments external to the corporation.

▶ ***JIT Backs Up into the Vendor's Operation.*** The classical approach to vendor relations was to do all the design and development work in house and then fix the process and freeze the specifications. The purchasing agent (PA) then called in the vendors, gave them a set of prints, and purchased parts on the basis of price and delivery, with the understanding that the quality levels had to be met. We find even now that companies well advanced in CIM technology and that have adopted the JIT methods in their plant still do not treat their vendors as part of their manufacturing operation. Some profess having JIT when in reality what they have done is to push all the fabrication problems back onto the vendors and to do nothing but assembly. When things go wrong, they punish the vendor by changing the vendor, but somehow they never seem to solve their problems. They just go on changing vendors and rejecting parts.

A more reasonable approach is to bring selected vendors into the product and process design phase of the operation. First, educate the vendor on what the problem is, then give them the technical knowledge and other soft tools to do the job, and jointly work out the best methods. Getting to know the vendor at this level will ensure that the price is right. An obvious extension of this thinking is to tie the vendor's computer data base into the corporation's. If production planning is talking about a model change in 3 to 6 months and plans are being formulated, the vendor should be in on them.

▶ ***There Will Be Fewer Vendors.*** Some large companies, such as GM, are heading toward just this. A "contest" is in progress now in which GM is saying that for a given item they will be using only a limited number of vendors, say three. Once the three have been selected, they will be privy to all information and will be told things that are not usually shared with vendors, such as specific market forecasts and engineering design information. Their computers will be directly connected to the GM MAP network. GM also goes to their vendor plants to check equipment, set up quality assurance programs, instruct workers, and the like. Figure 8.4 shows how the vendors might be tied to the user's host computer network.

8.8 Automated Warehouses

▶ ***The Warehouse Is Part of JIT Operations.*** Today, the automated warehouse, if a company has one, is simply another island of automation. Advanced data-entry and processing devices like the bar coder are moving into the warehouse rapidly and will accelerate the

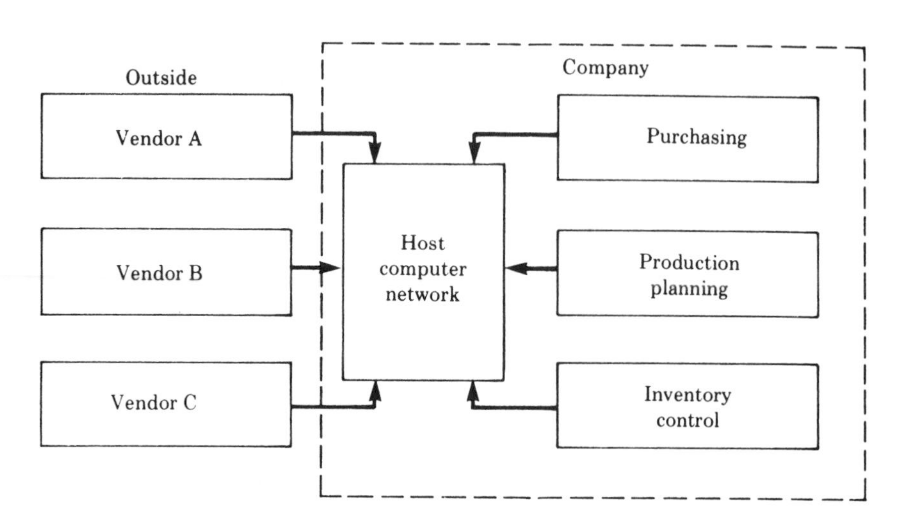

▶ ***Figure 8.4*** Integration of Vendors

automation process. A very large segment of the robotics industry makes a variety of automated warehouse equipment and automated material-handling equipment. Automated warehouses are an extension of the MRP method of operation that find wide support in industry.

▶ ***MRP Keeps Track of What Is Needed.*** Materials requirement planning (MRP) is a straightforward concept in that individual part use is determined weekly or monthly. Part availability is also noted on the same basis and a part or material balance is taken each period. If it looks like parts will be needed at any given time, the part order can be scheduled by factoring in the lead time. The same procedure is used by accountants and others in making cash forecasts, only the part in this case is money. Parts MRP is somewhat more complicated because many different parts make up an assembly and each must be accounted for. Also, many of the same parts are used on different assemblies (e.g., screws and other hardware items). The production planning people can determine inventory levels, floor space, and personnel requirements along with the order levels and order frequencies. Since it is an exercise in number crunching, it lends itself to the computer for solution. A simple example of MRP is given in Figure 8.5. Start with the production forecast and work backward to the time when you need to place the order.

Since each assembly is made of several parts, a detailed parts list is required for each assembly. From this, the individual part MRP is obtained. With a large assembly such as an appliance, the parts breakdown can number in the hundreds or thousands. The MRP program then becomes too large and complex to be handled by anything except a computer. MRP works well if kept simple.

Part \ Week	1	2	3	4	5
On hand (beginning)	26	62	32	55	41
Usage (during)	14	30	27	14	20
Received (during)	50[1]	0	50	0	0
On hand (ending)	62	32	55	41	21
Ordered (2-week lead)	50	0	0	50[2]	0

[1] Ordered 2 weeks ago
[2] For delivery in 6th week

▶ ***Figure 8.5*** A Simple MRP for One Part

An offshoot of MRP is MRP II. Here the MRP stands for manufacturing resource planning. Notice the unfortunate use of the same acronym for two different concepts. This has created great confusion in industry. (And, yes, there is an MRP III.) Manufacturing resource planning encompasses all the ideas of materials requirement planning but expands on it by incorporating the materials requirements into shop floor control, production scheduling, machine loading, labor utilization, finances, and all the other things that go into a totally integrated planning and scheduling function. Since MRP II is much more complex than MRP, it has been much harder to implement and to date has not achieved the level of acceptance or support its advocates had hoped.

However, there is a paradox about MRP and the JIT-CIM operation. If the goal to eliminate inventory with JIT techniques in an integrated factory operation is achieved, resulting in the product being made available as it is needed, and if we incorporate and integrate our vendor's operation into our own, then it follows that MRP is an unnecessary technique. It operates on an operand that doesn't exist. The likelihood of a perfect CIM operation that eliminates all inventory is extremely remote, so there will probably always be a place for MRP. It is a very valuable tool.

8.9 Nontechnical Aspects

▶ ***Beware of Technological Overkill.*** It would be good at this point to take a look at our situation from a broad perspective. The natural tendency for engineers and other technical people is to focus on a problem and try to solve that problem with technological answers. Many times this is the proper approach, but many times it isn't. The technological solution is appealing and seductive. Its appeal is that it allows engineers to exercise and display their technical prowess. It is seductive because the engineer, who can be likened to a high-priced physician, is the expert on the subject, and we all generally tend to follow the physician's advice with little question as to the foundation or basis for the recommendation. Once the decision is made, unless something catastrophic occurs, we have to live with it. On the shop floor this means that we try to make that new piece of equipment work. We will never know if the machine we did not choose would have been better, and the technologist can usually easily allay our doubts by telling us how much worse things would be if we had, indeed, selected the alternative machine.

► ***Robots in Japan Are a Personnel Tool.*** The natural tendency then is to leap into the solution of a problem with a technological solution. Usually, the more expensive and complex it is, the better. This frequently leads us down the wrong path. One thing the Japanese have been trying to teach us, but which we have not been paying much attention to, is how to increase our productivity using inexpensive practices. The Japanese study a situation and implement many nontechnical innovations and practices before they use robotics, automation, complex computer networks, and other high-tech solutions. In fact, the Japanese do not consider robots and automation as a means to replace direct labor, but a means to allow their direct labor to operate in a safer, cleaner, more pleasant, and more productive environment. It is this management attitude that has allowed the Japanese to automate their factories with the support of labor. In other words, they do not consider robots and automation to be a technological solution, but rather a management solution. They have not hesitated to use robots and are extremely proficient in the practice. However, they only use robots where robots are the most logical choice. The robots are accommodated to the factory practices, not the other way around.

► ***Make Only What Is Required.*** It is important, therefore, to look at some of the nontechnological things we can do to increase productivity. Throughout this book we have stressed the idea that one reason for CIM is that it fits into the concept of JIT. It is possible, and usually desirable, to incorporate JIT procedures and techniques first and then to look at ways to fit CIM into the scheme of things. The driving force behind the use of JIT is the need to eliminate waste. Waste is anything, including manpower and materials, that is not being used to produce exactly what is needed to satisfy the immediate requirement. If we make more than we need right now, then this is waste. Therefore, *inventory is waste and* *should be eliminated.* The goal is to operate without any inventory whatsoever. We must also operate so as to be able to satisfy a customer's immediate need, and so shortages and stock-outs are equally wasteful. Once we understand this concept, the rest of the ideas fall into place.

▶ ***No Inventory Forces Management Changes.*** Operating without any inventory means we cannot build up reserves to take into account contingency situations. The entire system is driven by what the customer wants and when. The problem is that we do not know what the customer specifically wants (model, size, color, etc.) and when he or she wants it. This poses the dilemma of making things on demand, without waste, but without knowing what the demand really is. The Japanese have resolved this dilemma by a total alteration of methodology. A large part of the methodology is how they respect, treat, and motivate people, which we will explore in Chapter 12.

8.10 Plant Layout and Sizing

▶ ***Smaller Plants Are Possible.*** One thing we can do is place all the machines closer together. We can do this because we have eliminated the WIP inventory. Figure 8.6 shows this idea and uses the idea of the cell approach discussed earlier. There is no need to provide an extensive storage area by the machine because we will not be storing anything there, or at most production for one day. By eliminating WIP, we also eliminate warehouses and material storage areas. Without warehouses and storage, we can eliminate all the nonessential warehouse people and many of the material transporters. We can thus see that the investment in buildings is going to be smaller because we can operate with much less space. Utilities will also be cheaper because of a lower heat and light power load.

▶ ***Network of Small, One-Product Factories.*** The management of a multiproduct factory becomes exponentially more complex as the size and diversity of the product line increases. There is an optimum point where the marginal cost of the management of the factory equals the marginal return of the number of products produced. As we will see later, the actual production control function in the Japanese factory is handled by the workers themselves. Some U.S. factories, such as Lincoln Electric in Cleveland, have started these practices as well. The result is a strong force toward building a network of factories focused on a specific group of products, rather than one huge factory. The Japanese also tend to pick a specific market and concentrate their resources and technology toward satisfying it. Within the market they select, they will offer a very large variety of models, styles, and other customer features to satisfy that particular market segment.

Raw material warehouse | Work in process (WIP) | Finished goods

Turn | Mill | Drill | Grind

A. United States

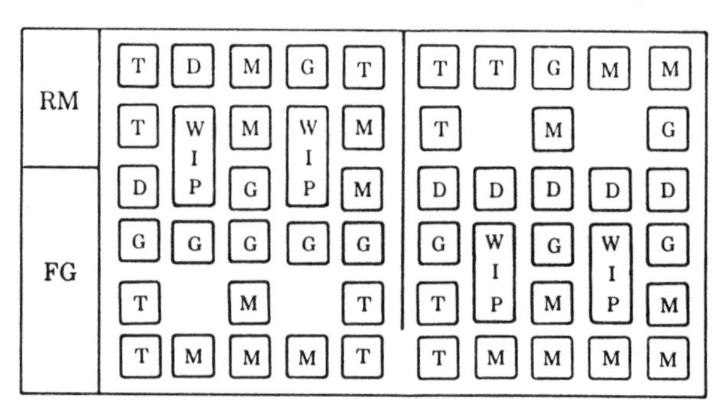

B. Japan

▶ ***Figure 8.6*** Plant Layouts

This leads to the idea that factories, even though extremely flexible in relation to options, are essentially making products that all fit into a single group. The mid-sized cars from factory X may be available in left- or right-hand drive, with or without air conditioning, and the like, but they will all use the same size frame and other major components. This form of group technology allows the combinations and permutations of a part to be made on the same machine and assemblies in the same factory. Raw material stock, tooling, set up, and processes will be altered to accommodate the specific design, but the basic machine and factory will stay the same.

8.11 Perfect Parts

▶ ***The Operator Is the QC Inspector.*** If a part being fabricated is to be used immediately upon completion and it is vital to the continuity of the production process, then it must be a perfect part. Perfect in this context means it meets the specification. Since we are not allowed to make parts for inventory and since we are not allowed to rework scrap (there isn't any scrap) we need to insititute procedures to ensure that every part is a perfect part. At least two things can be done to attain this condition. We make the machine operator the QC inspector and give him or her the necessary tools to do this. (Later we will discuss how we can make the operator want to be his or her own QC inspector, but assume for now that this is so). To be a QC inspector, we have to teach the worker all about QC. Much time and effort are spent in Japanese factories teaching individual workers the quality-control methods introduced into Japanese factories by W. E. Deming. We must also teach the operator how to apply these methods to the specific set of circumstances at the machines for which that operator is responsible. Many workers are proficient enough to be able to design their own control charts for their own processes.

▶ ***Use Coordinate Measuring Machines.*** In addition to operator knowledge, the machines should be equipped with the advanced automatic gauging to allow the machinist to see if the process is in control. The practice in this country is for the machinist to stop the machine and measure the dimension with a micrometer or dial indicator. Sometimes the machinist even has to remove the workpiece. One reason why this is so prevalent in the United States is the absurd practice of the machinist owning his own gauging tools. Some U.S. machine shops will provide or pay for calibration services, but don't bet any money that the machinist's micrometer, which is usually kept warm and handy in the back pocket, is capable of gauging anything precisely. Unless that machinist is a true and caring craftsman he will own the least costly tool available.

At the other end of the scale, both in sophistication and price, we find automatic gauging. Automatic dimensional gauging or CMM (coordinate measuring machines) mounted on the machine, which will shut off the machine when tolerances are exceeded, is the minimum that should be provided by management. Better still is gauging that is capable of controlling the operation of the machine through a servo loop.

8.12 Kanban

▶ ***Kanban Is a Simple System.*** One ingredient in JIT production is the use of the Kanban system. Kanban means card. A typical Kanban card is shown in Figure 8.7. The next time you go into a short-order restaurant, notice how the waitress takes your order on a piece of paper and gives the paper to the cook. The cook, who has a queue of these slips, prepares the order in due time and gives the food and slip back to the waitress. This is a Kanban system and this is the essence of a JIT

<table>
<tr><td>XYZ Corp.
Toronto, Ont.</td><td colspan="2">Engine Mfg.
Drilling x 14</td><td colspan="2">Part number
2Z43</td></tr>
<tr><td colspan="4">Part name:
Wrist pin 10 HP standard
Spec: 3 x PDQ: 26-36-26</td><td rowspan="2">Quant.

3</td></tr>
<tr><td>Provider:
Turning — cone 3</td><td colspan="3">User:
Grinding — head 36</td></tr>
</table>

Minimum information:

Part number/name
Quantity
Who is to do the work
What is previous operation

Drilling needs 3 pins and gives card to turning. Turning returns pins with card to signify order is complete.

▶ ***Figure 8.7*** Typical Kanban Card

operation. The fast-food shift operator doesn't get directly involved in the process nor do the managers down at franchise headquarters. The customer gets fed without any production control or WIP. The Kanban need not be a card; almost any signaling system will do if it indicates to the fabricator that it is time to make another part and what kind of part is needed.

The Kanban system can be easily changed to accommodate nonstandard situations. For example, some parts must be fabricated in batch sizes that are greater than the minimum lot. The Kanban card can be attached to the second box of a three-box stack. When the second box is empty, it goes back to the producer, who can then fabricate three more boxes. By the time the third box is empty, the producer will have made the three replacement boxes, again to arrive just-in-time. Here we do have a small amount of inventory. Realistically, a half-day or a day's worth of production is a sought-after goal.

▶ ***Balance the Plant According to Daily Requirements.*** Note in the three-box example that there may be several permutations of parts depending on what has been sold. If the day's requirements call for 2 blues, 3 reds, and 5 yellows, then this is what should be produced. In the United States we may produce 20 blues at a time, use 2 for the daily run and inventory the remaining 18. We do this to keep the machines running at capacity all the time. In fact, what we really do is pace our plant so that machines will run at capacity and keep our personnel on their machines to ensure that the lines are balanced to equal the rate of the fastest machine. This, of course, generates inventory. The Japanese,

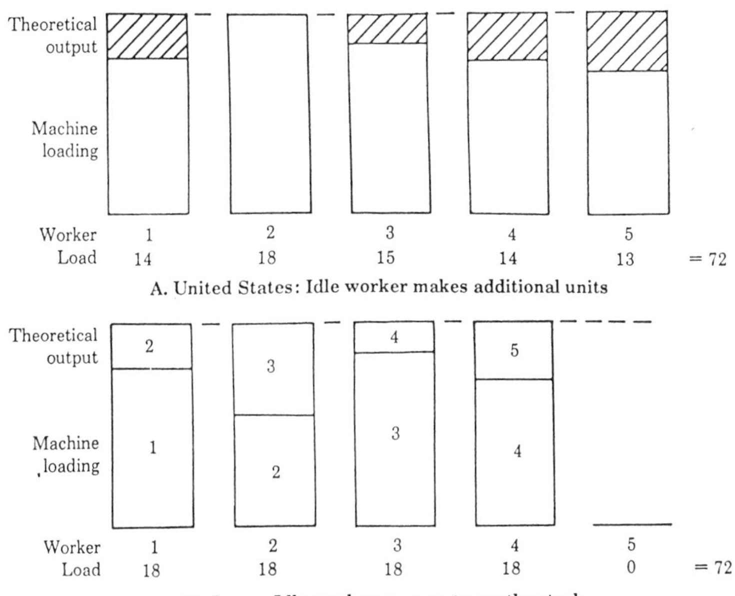

▶ ***Figure 8.8*** Japanese versus United States Machine Loading

on the other hand, balance the plant on the basis of the daily requirement. Because the Japanese worker is skilled in the operation of many machines, when the allotted production of the fastest machine is achieved, the worker is pulled off that machine and put onto a slower machine. The net result is that no extra parts are produced. This is shown in Figure 8.8. The discerning student will note that under the Japanese system the loading is adjusted by adjusting the labor input, and there will be less labor input because no labor goes into making inventory. When things get very busy due to increased demand in a particular department, another worker will be brought in from another department. This is only possible in an environment where worker functions are totally interchangeable.

8.13 Production Guidelines

▶ ***A Lot Size of One Is the Objective.*** The theory is great, but it cannot work unless it becomes economical to produce in very small batch sizes, with a lot size of one being the goal. However, we have learned from our operations management courses that production lot size is dependent on machine setup time. Thus the real goal is to minimize setup time. This is the major operational goal in the Japanese factory. And their success in this regard has been miraculous. In one study at a Toyota factory, the average setup time was over 60 minutes for 90% of the setups. Three years later, the average setup time had been reduced to less than 10 minutes for 90% of the setups. This is a better than sixfold productivity gain. In the United States it is usual for a press setup time to be 6 hours on a car body part. In a Japanese factory the same setup will take about 10 minutes. Because the economic lot size laws apply in both cases, the Japanese can achieve their goal of 1 day's worth of production with three setups per day, where we operate with a lot size of 10 days and only one setup per day. This figures out to an average WIP in the United States of 5 days, contrasted with 1/2 day in Japan. Here the productivity gain is tenfold.

▶ ***We Need to Reduce Waiting Time.*** The drastic reduction in lead time can be achieved in several ways. In U.S. plants we typically spend 95% of the time on a job waiting and 5% actually cutting metal, as shown in Figure 8.9. So we need to focus on the waiting time. Downtime on the machine comes from both internal and external causes. An external cause might be having to go to the tool crib to get the proper tool or waiting for the parts to be delivered from the prior operation. Having the proper tools at the site along with the parts and other local staging activities will eliminate a great deal of the external time. Roller carriers and roller platforms (Figure 8.10) to facilitate the loading and unloading of machines are widely used in Japanese factories. We want to eliminate lifting because it is dangerous and fatiguing, as well as time consuming.

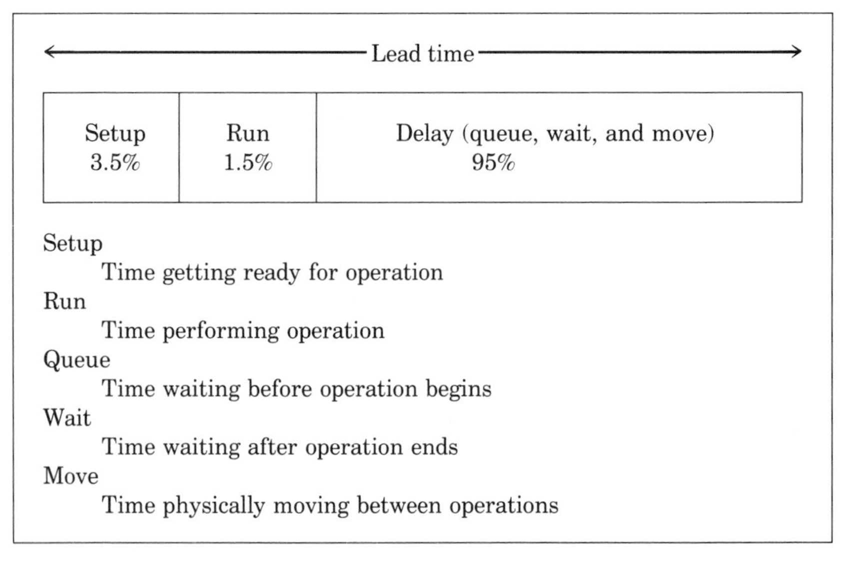

▶ ***Figure 8.9*** Components of Lead Time (*Iron Age*, March 1985)

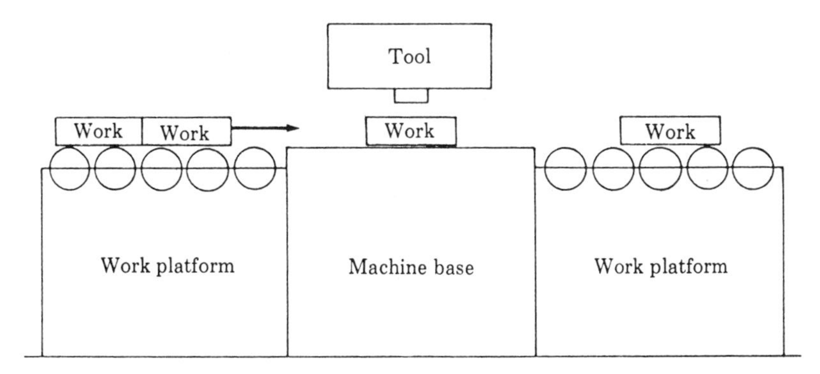

A. Roller platform in Japan

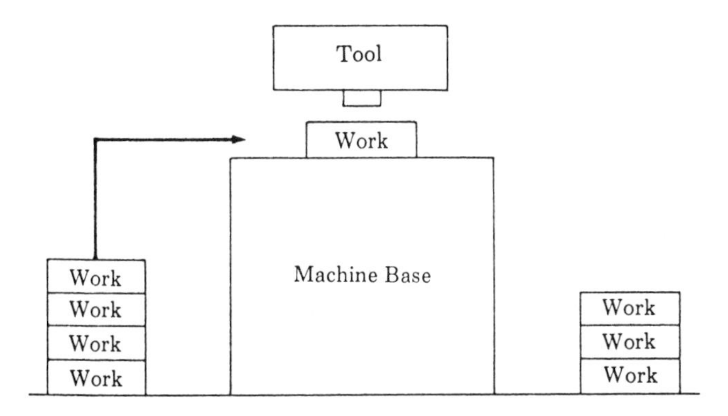

B. Lift method in United States

▶ ***Figure 8.10*** Japanese versus U.S. Work Transport

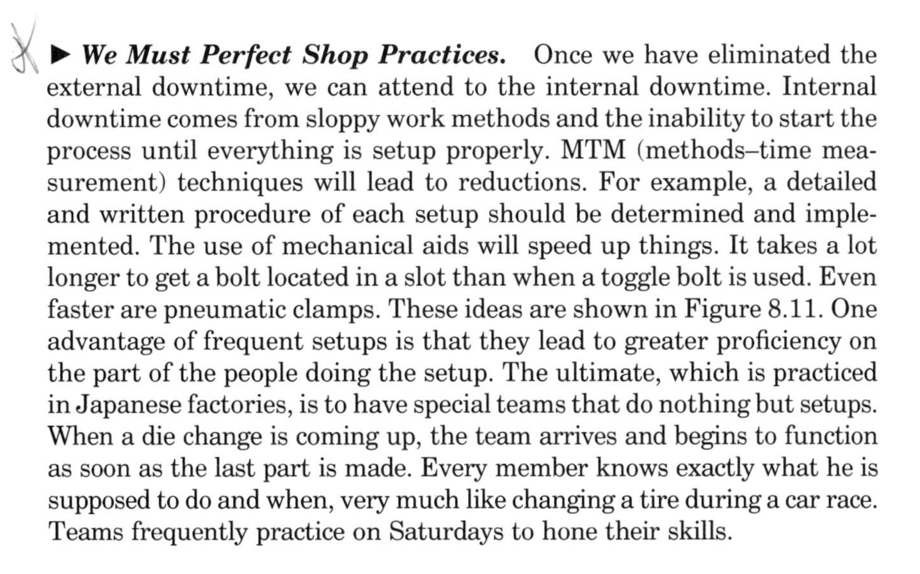

▶ ***We Must Perfect Shop Practices.*** Once we have eliminated the external downtime, we can attend to the internal downtime. Internal downtime comes from sloppy work methods and the inability to start the process until everything is setup properly. MTM (methods–time measurement) techniques will lead to reductions. For example, a detailed and written procedure of each setup should be determined and implemented. The use of mechanical aids will speed up things. It takes a lot longer to get a bolt located in a slot than when a toggle bolt is used. Even faster are pneumatic clamps. These ideas are shown in Figure 8.11. One advantage of frequent setups is that they lead to greater proficiency on the part of the people doing the setup. The ultimate, which is practiced in Japanese factories, is to have special teams that do nothing but setups. When a die change is coming up, the team arrives and begins to function as soon as the last part is made. Every member knows exactly what he is supposed to do and when, very much like changing a tire during a car race. Teams frequently practice on Saturdays to hone their skills.

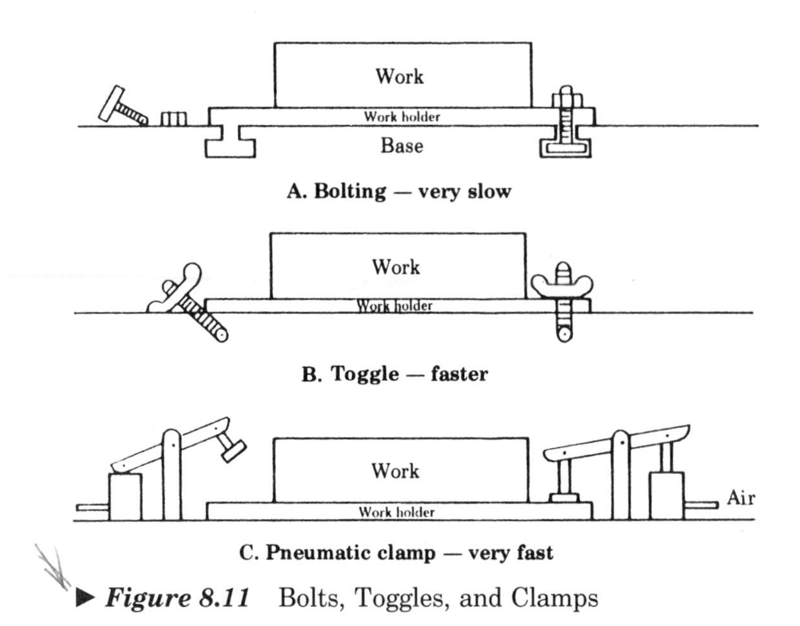

▶ ***Figure 8.11*** Bolts, Toggles, and Clamps

▶ ***Use Pre-setups and Locating Devices.*** Another way to reduce internal downtime is to eliminate adjustments once the piece and tools are on the machine. Copious use of locating pins, stops, and positioners is warranted (see Figure 8.12). If all the molds have the same size base, all have the same thicknesses, and the locating pins on the molds and the holes on the platen are in the same place, then changing from one mold to the other in an injection molding machine becomes quite rapid. Preheating the die-casting mold offline with external heaters means you don't have to wait for the mold to heat up on the machine. A goal often achieved in Japanese factories is that the first part made after a changeover is a good part. Even a simple thing like buying extra tool holders and mounting the tool off the machine can increase productivity dramatically.

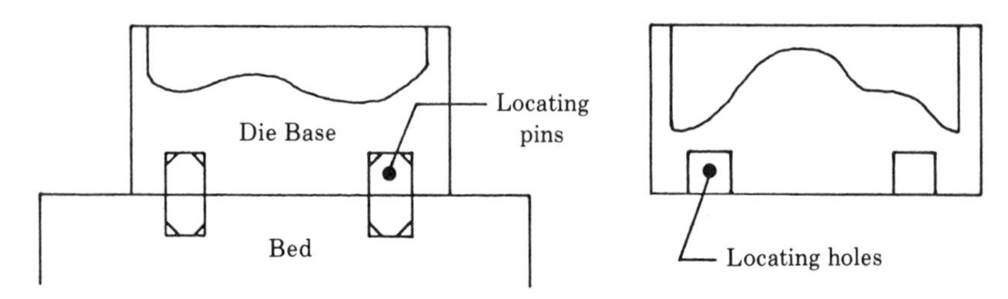

All die bases are the same size. All pins and holes are in the same location.

▶ ***Figure 8.12*** Use of Locating Pins to Speed Setup

Implementing JIT

Looking at the entire operation and trying to figure out how to implement JIT may appear to be an overwhelming job because of its size. Fortunately, we can approach the problem by looking at each process and making three key ratio measurements to see if a particular operation is favorable or unfavorable. In all cases the ratios should ideally be 1 to 1, with a 2-to-1 ratio being considered good. Anything higher than 3 to 1 should be considered unfavorable. After we have measured each process, we can then rank them and turn our attention to those processes that have the worst ratios.

The first ratio is lead time to work content. If a bin holds 40 parts and it takes 12 minutes to mill a slot on each part, then it will take us 12 × 40 = 480 minutes to work our way through the bin. Since each part takes 12 minutes, we have a 40-to-1 ratio of elapsed time to actual work time.

The second ratio is process speed to consumption rate. If a process can cut 20 pieces per hour and is feeding an assembly operation that can handle 5 parts per hour, we have a ratio of 4 to 1. This means that the cutter is only working 15 minutes in each hour or it is building up inventory by running all the time.

The third ratio is number of pieces to number of stations/operators in the production line. If we have 100 pieces in the production line and 10 people doing something like attaching another part, we have a ratio of 10 to 1. In this case, 90 parts are WIP. The solution is to institute a Kanban system so that we can bring the parts-to-people ratio down to a reasonable level.

The procedures and ideas presented here are a small sample of the kinds of things that can and ought to be done to increase productivity before we attempt to use complex and high-cost automation to solve the problem. "Simplification before automation!" should be the motto. However, we frequently read in trade magazines about how someone's robotized, automated, and chrome-plated assembly line has failed to achieve operational goals or is X number of months behind schedule. One large firm, ready to use a whole series of robots, did a study to see how they would fit them in. As a result of the investigation, they instead changed their work practices and concluded they didn't need any robots at all.

8.14 Noise

▶ ***Noise Causes Problems.*** In a manufacturing environment, the air and wires are filled with radio-frequency (RF) fields and transients. This is called noise. Whenever a switch or relay contact opens or closes, a transient spike appears on the line and RF energy is generated. Early computer devices and controllers were not transient protected and interesting situations sometimes developed, such as heaters on line 2 turning on when line 1 was started. As computers have moved out of their air-conditioned and electrically isolated offices and onto the shop floor, more attention has been given to the noise and transient problems. Even so, whenever a glitch or malfunction occurs, you should ask if the problem is line transient or RF noise before looking into the computers or controls. Such action may save many dollars and much frustration.

► ***Noise Is a Production Killer.*** No matter how well a system is designed, it will not operate if electrical noise interferes with the operation of the computer function. The noise problem is really a subset of the overall need to make the robotic device factory-hardened. Machine tool builders are well aware of the need to make their tools rugged and able to operate in environments where there is gritty dust, acid fumes, solvent vapors, excessive heat, humidity, or excessive cold. These are conditions found in some of our heavy industries. A steel mill or foundry is typical. The situation regarding noise requires similar special considerations. Many people who build programmable devices, particularly machines designed to operate in the lab or in an air-conditioned computer room, have not designed their equipment to operate in the harsh factory environment.

► ***Noise Is Pervasive.*** Electrical noise is found throughout the factory in both the electric service wires and the air. Everytime a motor starts or a relay closes a very large transient pulse is generated, which gets out into the air and onto the lines. Other sources of noise include circuit breakers, commutator brushes, and anything that generates a magnetic field. These pulses are directly passed on to the delicate computer circuits either through the wires or by induction. Interference with data flow is only one problem that can be encountered, and in severe cases great harm can be done to the equipment. Particularly insidious is noise interference on an intermittent basis because it is difficult to detect. If your fancy new robot isn't doing what it is supposed to, make sure the problem isn't caused by noise in your plant before you call the manufacturer. Most electricians are not too sensitive about noise and will frequently run power cables in the same conduit with signal cables. Check this out first.

► ***Use Shielding and Filtering.*** It doesn't take much to cause a problem. A typical TTL circuit operates at 2 nanoseconds, which is about 500 megahertz. Circuits can detect noise up to about 10 ns, which means the equivalent wavelength is about 0.03 meter. A small slit or opening in an enclosure one-quarter this long, about 3 inches, will allow whatever RF energy hits it to pass into the enclosure. The case has to be tightly closed and all openings should be covered with screening.

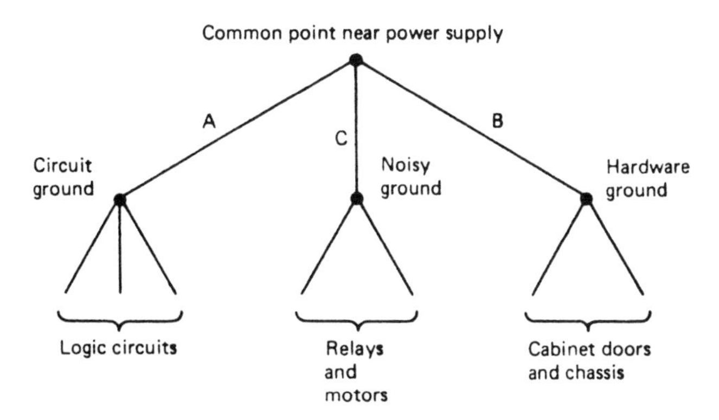

▶ ***Figure 8.13*** Grounding Tree (Snyder, *Industrial Robots*; Copyright by Motorola, Inc.)

Power-line noise can be filtered out with various low-pass filters and bypass capacitors. In extreme cases, special isolation transformers may be required. It is fundamental that all equipment be grounded with respect to a neutral return. Figure 8.13 shows a grounding tree. Circuits are grounded to a common point, motor bases are likewise grounded, and so is the door and cabinet. You cannot rely on earth ground as there may be a large potential difference between one earth ground and another. Significant currents will flow.

▶ ***Need for Good Grounding.*** Noise picked up by induction will create current ground loops, and so all cabling must be shielded and properly grounded. Grounding should take place only at the receiving end of a signal cable. This will prevent excessive currents from flowing in the shielded portion of the cable because of unequal earth grounds. Many times adjacent wires will pick up signals from each other, causing cross talk. Using twisted wires in the shielded cable will help.

▶ ***Impedance Matching Helps.*** Data transmitted at high speed need to be placed on high-frequency coaxial transmission lines such as are found in cable TV. If the lines are not at the proper impedances and if the loads are not matched, reflections will appear and will interfere with the signals present on the lines. A balanced circuit is shown in Figure 8.14. Notice the coaxial cable is grounded only at the receiver end. This prevents ground loops. However, the solution to the reflections problem is usually best left to network experts.

▶ ***Look Inside the Enclosure.*** Noise inside the enclosure can be generated by differences in the switching of the transistor circuits. In some cases, optical isolators (Figure 8.15) may be required, and the problem may be in the original design. Other design considerations

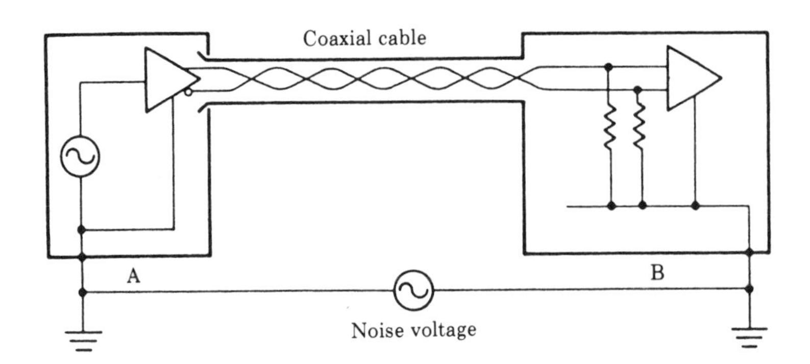

▶ ***Figure 8.14*** Balanced Driver and Receiver Pair (Snyder, *Industrial Robots*

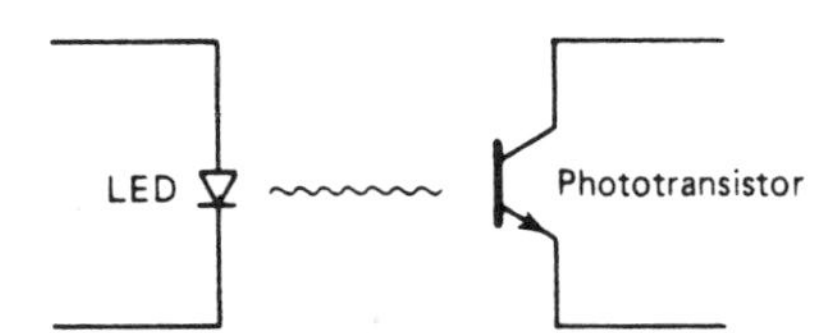

▶ ***Figure 8.15*** Optical Isolator (Snyder, *Industrial Robots*)

would be to ground every other wire in ribbon cable and lay out the board so that there are short runs between active circuits. Figure 8.16 shows these ideas. There are many possible solutions. Although these are vendor equipment problems, you may have to solve them or be able to point out the deficiences to the vendor.

If noise problems are suspected, check with the manufacturer to find out what they have incorporated into their machine to prevent noise from entering their equipment. If you are sure they have taken everything into account and they have had successful installations in similar factories, the equipment was probably installed improperly in your factory. It is to be hoped that the problem can be solved by something simple, like tightening the cover bolts or plugging the improperly located mounting holes that you drilled.

8.15 Summary

We need to study the total system design if we are to successfully implement CIM. For CIM to be cost effective, we need to use modern factory methods such as JIT before we automate. Only by having a clear

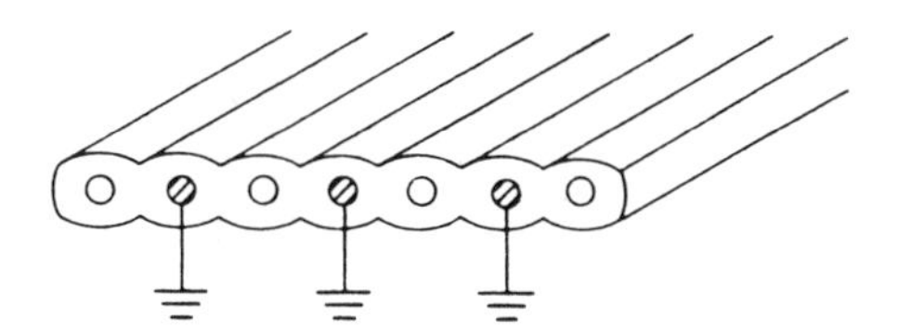

A. Use of ground in flat ribbon cable

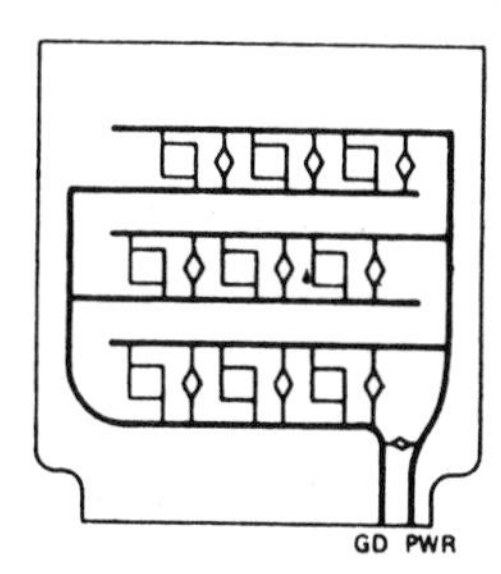

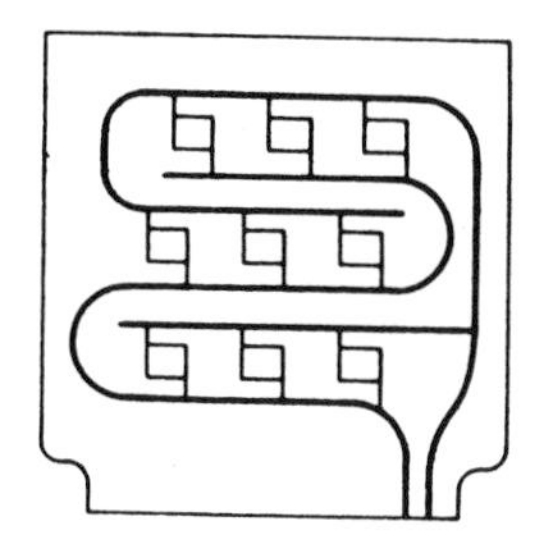

B. Proper circuit board layout **C. Improper circuit board layout**

▶ ***Figure 8.16*** Ground Handling to Reduce Noise (Snyder, *Industrial Robots*)

picture of the entire process and its various components can we begin to consider technological solutions.

One result of implementing JIT into the environment is that all the processes and people become much more closely coupled. Any equipment failure in a thoroughly integrated plant will create extreme financial and production problems. For this reason, a high degree of implementation of CIM may not be the best choice. Many large concerns are finding this out.

Many nontechnical things can be done to ease the production situation. Elimination of WIP by eliminating idle machine time and waiting queues is cost effective. So is reducing transport and lifting by people and the copious use of pre-setups, locating pins, and proper line balancing.

8.16 Exercises

1. Why is it important to have the systems designed?
2. What is the advantage of a just-in-time type of operation in terms of management?
3. How is a robot used in cell structure?
4. What does the term product line mean?

5. What is the operational goal in a just-in-time or flexible manufacturing plant in terms of batch sizes?
6. What do we mean by staging?
7. What changes are taking place in the relationship between purchasing and their vendors?
8. What kind of information will likely need to be shared with vendors in the future?
9. What kinds of planning need to be incorporated into an MRP program?
10. How does the Japanese attitude toward robots differ from the American?
11. What is the driving force behind just-in-time?
12. If we operate without inventory, how do we handle contingency situations?
13. Why can we have smaller plants?
14. Who handles the production control function in a Japanese factory?
15. Who does the quality control in the Japanese factory?
16. Why do we need to use more automatic dimensional gauging?
17. What does Kanban mean and how is it applied?
18. What is a realistic inventory level in a just-in-time factory?
19. How do Americans and Japanese balance their plants?
20. What things do the Japanese concentrate on in factory management?
21. Why is staging important and how can we facilitate staging?
22. How do we eliminate internal downtime?
23. Why are written procedures important?
24. Why do the Japanese have special teams for doing setups?
25. Why are locating pins, stops, and positioners used in Japanese factories?
26. What are some of the causes of electrical noise?
27. If you are having problems with noise in a factory, what is the first thing you should check out? Why?

9

Design of the Data Base

9.1 Introduction

Types of Data Generated

The central idea behind CIM, integrating the manufacturing operation with the computer, hinges on what the computer or programmable device is to do with what other PD and when. Data may be used as they are being generated. This is called real-time operation. Examples are: if a bearing sensor reads that a key bearing in a motor is overheating and this information is used to alert someone to do something; or a robot arm proximity sensor may discover something in the path of the arm. There must be provision to immediately stop the arm so no damage is done. However, by far the majority of data generated are stored somewhere for future use. The data must be accessible when needed and in a format that is intelligible to the PD accessing the data.

At other times we need to accumulate data, but we may not be interested in doing anything with the data immediately or we may want to accumulate data so that we can do an analysis on all the data. In this case we need to provide storage for the data and to develop means to properly manipulate or operate on the data. Furthermore, data of this sort are usually tied to other kinds of data being generated by the same machine or even machines in other locations. So we need some kind of dictionary or directory to allow us to tap into the right data for the task at hand. All this comes under the general topic of data-base management, the subject of this chapter.

9.2 GIGO and Sufficiency

▶ ***We Need Enough of the Right Kind of Information.*** GIGO is an acronym for computer jargon meaning "garbage in, garbage out." The data that go into the data base must be correct. Correct here means not only that the data elements themselves reflect true conditions, but also the data must be complete and unambiguous. The x315 distributor cap must be listed in the inventory as part number x315. If the warehouse worker counts 15 distributor caps and 7 of them are x315 and 8 are X315, it would be incorrect to punch into the computer 15 of x315. It may well be that the x315 and the X315 are the same part or that they are interchangeable, but how is the computer to know unless it is programmed correctly? The count must be entered properly.

Notice we have touched on three elements with this simple example. First, the data must be entered correctly. Second, the data must be in the proper format, and, third, the data must reflect reality. Entering data correctly can be facilitated by checking the work of the data-entry people. Also, it is a good idea to build a data audit trail into the system so that incorrect data can be found. The computer is fast but stupid and will do whatever it is told with whatever it has to work with. Constant vigilance is required to ensure correct data.

Assuming the data are correct, another problem can develop if the program is incorrect. We want to add units of x315 as we bring them into inventory and subtract them when we take them out. If we do it the other way around or only do it sometimes, we will end up with incorrect results. Programming is a key element. And GIGO is the watchword.

9.3 Integrating Software

▶ ***Integrate the Data Base First.*** Before we can integrate the software, we need to determine the layout of the data-base format in each data base within the company. Data bases evolve depending on the particular requirement of each operating department. It is unrealistic to assume that each department will need the same data base or even that all departments will access a central data base. This might work in a very small company or in a company just starting out, but it has not happened that way in larger companies. The reason is that traditionally the Management Information System department, usually controlled by the financial vice-president, provides data and information first to the people who are paying their salary, that is, the financial department. Other operating departments are put on a low priority status; as a result, they find other means of developing needed information independently. With the advent of the personal computer and the software to go with it, each department could inexpensively and in timely fashion develop their own

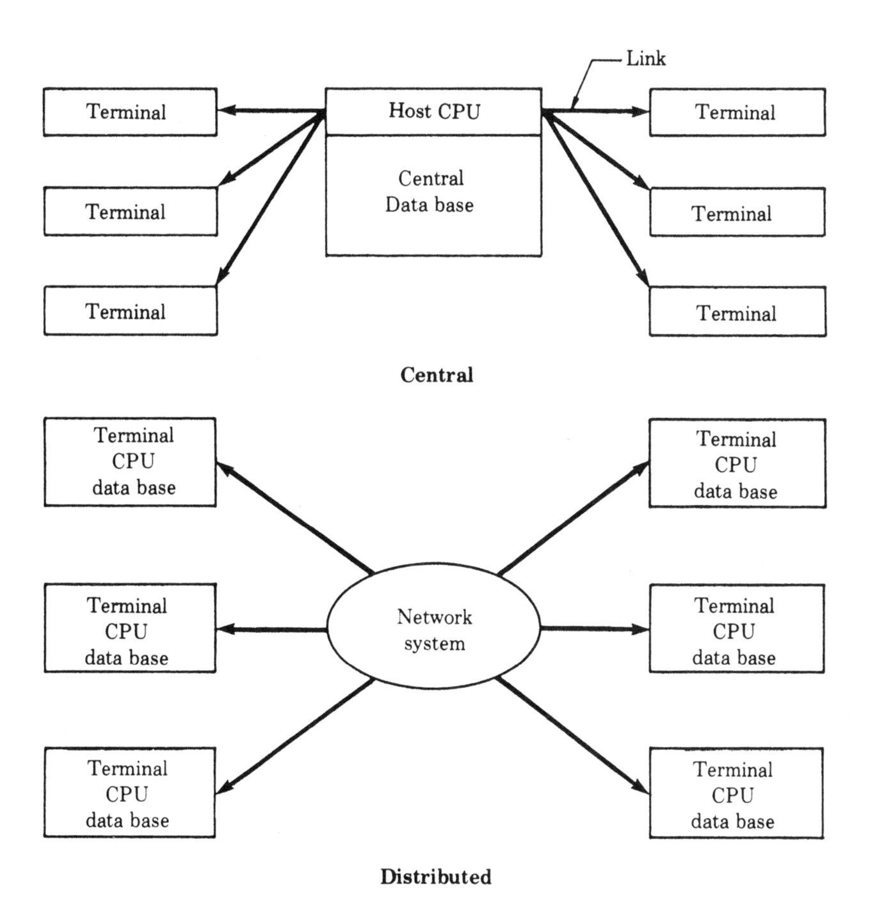

▶ ***Figure 9.1*** Central Data versus Distributed Data Base

data base and get it operating. This has created islands of data bases. Figure 9.1 shows a central versus a distributed data-base model.

Local Area Networks

A concerted effort is being made to integrate these islands of data bases by employing local area networks (LANS). But large technical problems must be resolved and this means large expenditures. The de facto solution to the situation has been to settle on a semistandard as far as software is concerned, and it has evolved that MS-DOS is now the unofficial operating system in most companies. When buyers say "IBM compatible" what they really mean is they want to operate on MS-DOS. Other manufacturers missed the point that the power of the MS-DOS system is that all the data on all the floppy disks are formatted the same way. Thus, if someone in Kokomo needs data, all the person in Muncie needs to do is to mail a copy of the data floppy disk. In the same office, people pass disks around as needed and seldom bother going through the MIS department or using the LAN. This is referred to as the "sneaker net." Two popular formats at the PC level are ASCII (American Standard Code for Information Interchange) and DIF (Data Interchange Format) created by VisiCalc. At the mainframe level, each manufacturer has its own system or variation.

Sequential Access

As we move up into the software programming level of data base design and management, we find there are two general ways of getting the data into the storage or memory. It is possible to combine these two ways in a variety of fashions to come up with a large number of permutations. One way is to store the information in sequential access file fashion. The term sequential as used here means that we place the data element groups or fields into memory in a stacking fashion, like loading trays in a cafeteria or ammunition in a magazine clip. A group of elements, which could be of any theoretical length, is placed into storage with place markers at the beginning and end. A typical marker might be a carriage return or a comma. To find a particular field, we have to go through the entire file until we come to what we are looking for. We do this by counting the markers. If the field we want is the 900th, we need to go through 899 fields to find it. The advantage is that we do not need to be concerned with field or size of the data element groups. Sequential access is handy in situations where we want to read the entire file, process it, and put it back into storage. Word processing is a good example of this kind of situation. Its advantage is that it takes the minimum amount of memory. Its disadvantage is that execution times become large with massive files. A sequential access system is depicted in Figure 9.2.

Random Access

The other type of file is the random access file (see Figure 9.3). As its name implies, we can find the data quickly if we know where to look. To

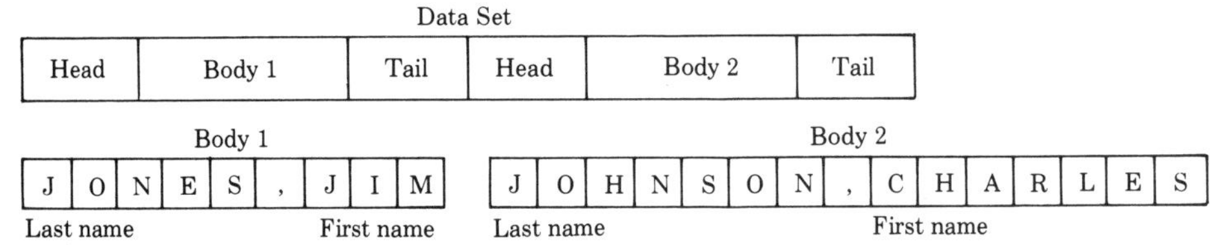

▶ ***Figure 9.2*** Sequential Access System

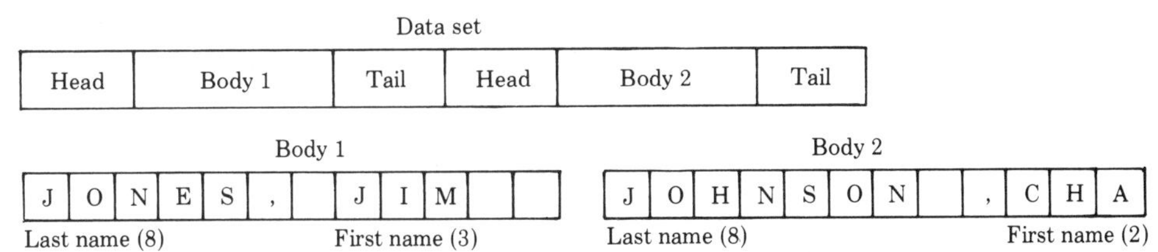

► ***Figure 9.3*** Random Access System

accomplish this, we assign a specific number of data elements into each field. Often we have fields that could contain more data elements, and sometimes we have more data elements than can fit into the field. As a result, we must be careful when we set up the data base to ensure that we do not lose information and do not waste space. To access the file, all we need to know is the address of the field. Knowing this, we can skip over all the other fields to get to the one we want. Here the trade-off is speed for memory. Random access is used in spread sheets, accounting records, and similar situations where it is easy to assign definite field lengths.

► ***Data-Base Management Relates Different Pieces of Data.*** The data may be in the storage locations in either sequential or random mode, but the data are of no practical use unless we can access, operate on, and then return to storage or display the data. If the data are relationally simple, meaning that all the data can be classed in a single or very few groups, we have no problem finding them. For example, a machine may be generating 5000 data points on the temperature of one bearing. If we want to know certain statistical information about the bearing, all we need do is recall the data, do the analysis, and provide the answer. In a more complex situation, the data may be intertwined or inextricably linked with other data items. Accessing one piece of data and doing something with it will have implications for the rest of the data. Consider what happens when a product is sold. We have product data, inventory data, parts data, sales data, cost data, and so on. Each category of data will have to be changed to reflect the fact that a single item of merchandise was sold and delivered to a customer. So we need to be concerned not only with the data in each field but also with the changes that will be effected in many other fields that have some sort of relationship to the data in question. This is the topic of data-base management.

Data Models

Two basic models are used in data-base management. These are the relational and the network or hierarchical models. Again, there are many variations on these two models and many different ways in which they can be used. Relational data models are generally used with small data bases and hierarchical data models are used with large data bases, but not always. In general, relational data are used with interactive programs such as might be found on a PC where the user needs to get immediate answers and wants to play "what if." The language structure is manipulative, and typically we find FORTRAN or BASIC types being used at this level. The hierarchical data model is used with large mainframes and employs languages such as COBOL; although the machine processing is usually faster, answers are not quick because of scheduling and other people considerations.

► ***The Relational Model.*** The relational model is structured on the idea of storing the data elements in cells, like the mail boxes in the post office. The inventory of different models of motors at various warehouses around the country lends itself to this cell or matrix structure. Column headers would be the warehouse locations and the row headers would be motor model number. The inventory of each part would be the data in the cell. We can manipulate these data in various ways. For example, if we want to know how many parts are in Seattle, we can call up that column, and if we want to know how many 1-horsepower motors there are, we can call up that row. We can complicate the picture very quickly by saying that all green motors sell for more than $100 and all red ones sell for $100 or less. We must now set up additional matrices showing the size of the motors and how many red and green ones there are. For the system to work, we need to establish relationships (hence the name) between the various matrices and operate according to strict set procedures. Note that the idea is to assign a cell address and move the data about by manipulating the addresses. Nearly all the popular PC data bases are relational.

Figure 9.4 shows a relational model for cars and Figure 9.5 shows the relationship between cars on order and the car dealers who ordered the cars. Figure 9.6 is a model relating the orders with the car and the dealer. In this example, we can determine that Manfred in Peoria is expecting Peccadillos worth $15,000 each.

► ***The Hierarchical Model.*** The hierarchical data-base model is structured so that certain additional informational data elements are attached to each group of data elements. The minimum information attached is an identification number for the cell and the ID number or pointer to the next cell. More advanced models will also have the ID number or pointer of the previous cell. In this way, data cells can be linked together in chain fashion. To find a particular cell of interest, we need to know the route but we are not concerned with the address. Hierarchical models are also known as network models because the cells can be thought of as the nodes in the network and the paths as the links.

Name	Class	Length	Cost
Zinger	Compact	8.3	7500
Spinner	Compact	8.1	7150
Cruiser	Sedan	9.4	8540
Archer	Sedan	9.6	9000
Bison	Pickup	12.4	12000
Peccadillo	Luxury	20.0	15000
Valhalla	Luxury	22.3	20000

ATTRIBUTES

TUPLES

VALUES OF COST ATTRIBUTE

► ***Figure 9.4*** A Sample Relation for Autos (Freiling, *Understanding Data Base Management*)

ON ORDER

Auto	Dealer	Qty	Date
Peccadillo	Manfred	5	3-10-81
Zinger	Manfred	2	3-12-81
Bison	Samson	3	3-15-81

DEALERS

DName	City	State
Manfred	Peoria	IL
Samson	Knoxville	TN
Atlas	Newport	ME

► ***Figure 9.5*** Sample Relations of Orders and Dealers (Freiling, *Understanding Data Base Management*)

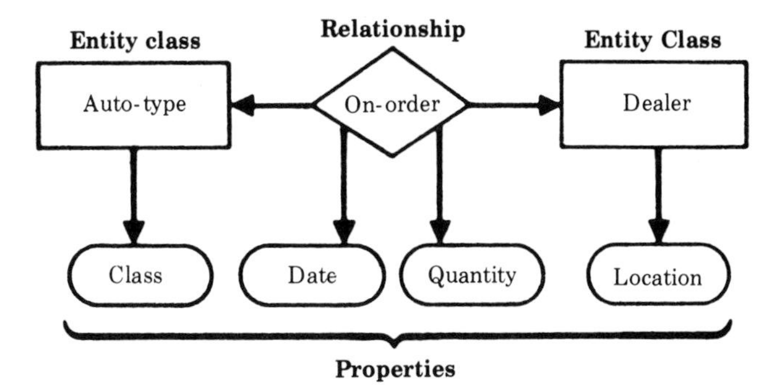

► ***Figure 9.6*** Relational Model (Freiling, *Understanding Data Base Management*)

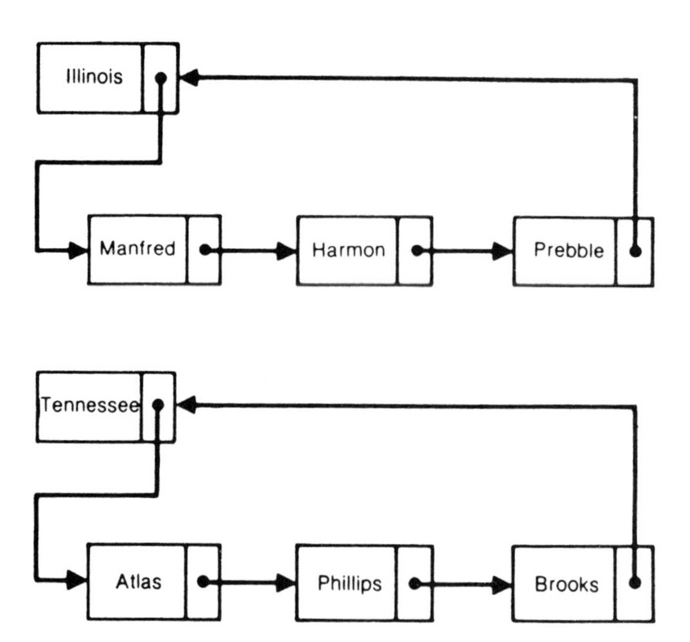

▶ ***Figure 9.7*** Data Sets (Freiling, *Understanding Data Base Management*)

Having both forward and backward pointers allows us to chain in either direction and also allows us to easily insert and delete cells. When this is done, we simply change the pointers in the adjacent cells to reflect the addition or deletion. In the example of the motors and distributors, if we want to know the number of motors in Seattle, we follow path A, and if we want to know how many 1-horsepower motors there are, we follow path B. Both paths need to be specified before we start out. There are explicit rules and procedures regarding what data can be in each cell, how the cells can be tied together, and what kinds of paths can be used. The hierarchical model can be thought of as being similar to a road map; the numbered roads are the links, the cities are the nodes or cells, and the population in each city is the data we are after.

In the car example, we see a data set for the dealers in Illinois and Tennessee in Figure 9.7. The data links for dealers, autos, and what is on order are shown in Figure 9.8. These links are called intersection records. Again, we see that Manfred is expecting five Peccadillos. Another data link, not shown, gives the path to the value of the Peccadillo.

9.4 Artificial Intelligence and Expert Systems

▶ ***AI Processes Much Data for a Single Answer.*** There comes a time when the amount of data we need to process in a given period of

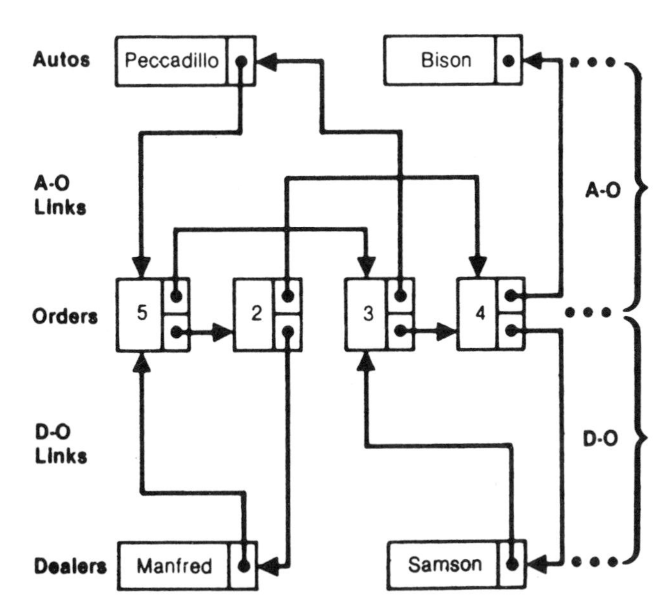

► ***Figure 9.8*** Intersection Records (Freiling, *Understanding Data Base Management*)

time becomes overwhelming or we don't get the answers soon enough. We find this situation in the integrated factory in several key areas, all having to do with artificial intelligence and expert systems. Useful products employing these technologies are just starting to appear in the commercial marketplace. Until very recently, all the equipment, algorithms, programs, and what not were the private domain of academia. Depending on who you read, expert systems is either a part of AI or it is a separate topic. In this text, expert systems are a part of AI. A central idea is to try and emulate the cognitive processes of the human brain and apply them to real-life problems.

Figure 9.9 is a model of the AI process. The AI system relies on knowledge or a data base, a means to search the data in an efficient way, and the ability to deduce valid conclusions. The central problem is to get a few conclusions or answers from a great deal of data and get the answers quickly. Vigorous research is going on to design easily accessible data bases and efficient search algorithms and to come up with meaningful cognition programs.

► ***There Are Few Commercial Applications of AI to Date.*** Sinceapplication has more marketability than emulation, there are several distinct applications available now that to a greater or lesser extent use the knowledge gained from the AI community. A great deal of research is continuing and the whole area of AI is exciting. From a practical point of view, we find AI is used to process vision information for feedback

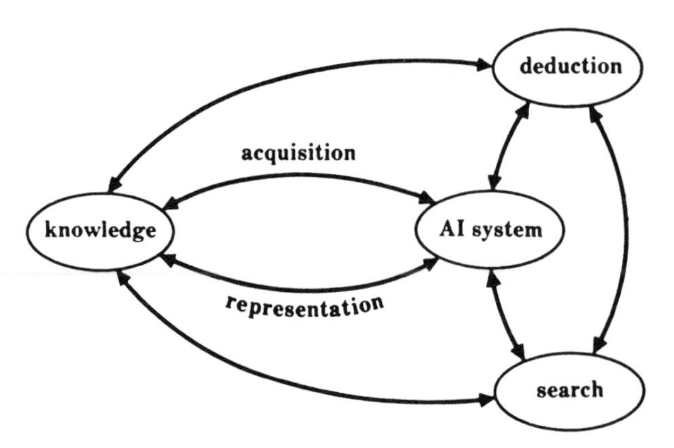

► ***Figure 9.9*** Model of AI System (Gloess, *Understanding Artificial Intelligence*)

control of robots (Figure 9.10), for speech recognition for data entry (Figure 9.11), and to handle planning functions for process control and warehousing, an expert system application (Figure 9.12). Future applications will be to make robots and other automation devices able to learn and apply knowledge in a commonsense fashion, to develop expectations and anticipate events, and to adapt to a changing environment. At present, there are few expert systems on the factory floor. Most expert systems are in areas such as medical diagnostics, law case and review, and bio/genetic/chemistry applications, where judgment from extensive training and experience can be used.

► ***Processing AI Information Is Slow.*** Most current AI applications involve a great deal of local data storage and immediate number crunching. In the case of vision systems, each pixel needs to be read, assigned a relative intensity on the gray scale, and digitized. This image is then processed to reveal the significant points, such as edges, planes, and other geometric considerations. In some cases it is desirable to rotate the picture and alter it so that perspective is eliminated and either a plane or elevation view is presented. This view, with the important geometric details emphasized, may then be compared with views of objects stored in memory for comparison. Factors such as differences in magnification, rotation, and proximity to other objects between the unknown and standard need to be resolved so that a valid comparison can be made. Specifications as to what constitutes a good object can then be applied to accept or reject the object. There are good machines that can do these things in a static situation or on slowly moving lines. The trick is to do it at high speeds. The problem then becomes one of data storage capacity and processing speed.

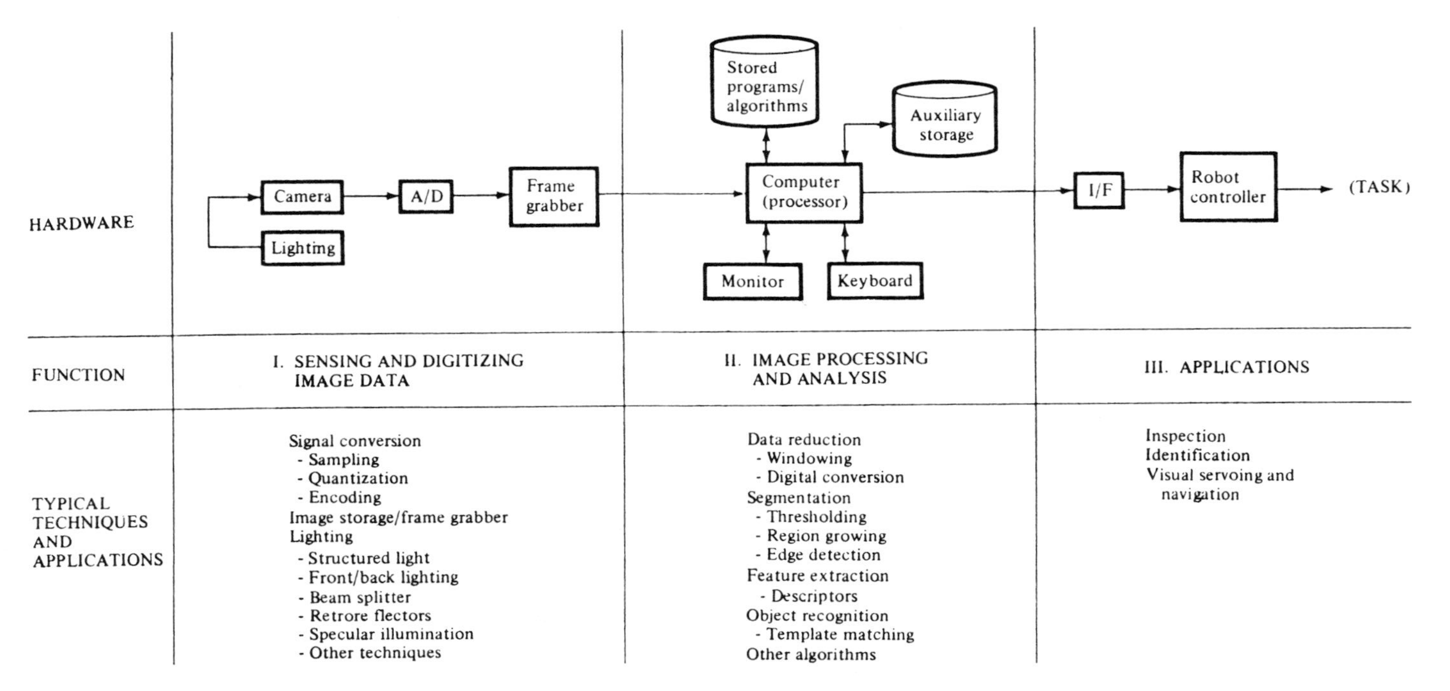

Functions of a machine vision system

▶ ***Figure 9.10*** Functions of a Machine Vision System (Groover, *Industrial Robots*)

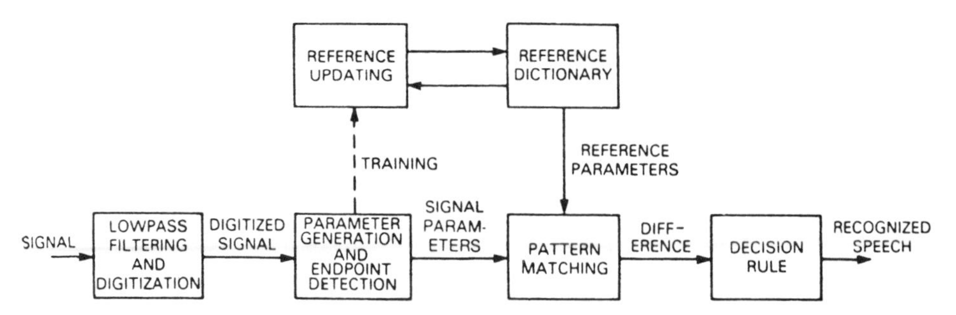

▶ ***Figure 9.11*** Basic Speech Recognition Paradigm (Zue, *Natural Languages, Speech,* Part II)

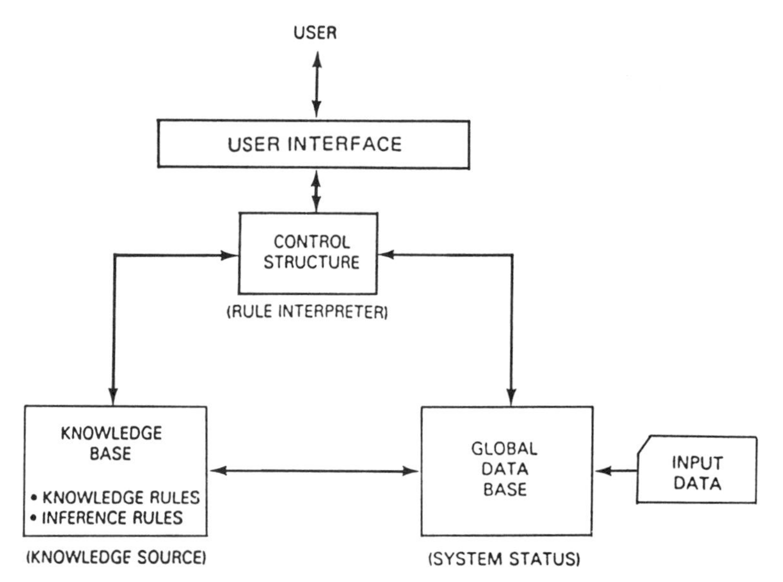

▶ ***Figure 9.12*** Basic Structure of an Expert System (Gevarter, *Intelligent Machines*)

Parallel Processing

There are two ways to speed up the process. One is to build faster sequential machines. A supercomputer such as the Cray machine is very fast and has a large memory. It sells for millions of dollars and is too expensive for the typical machining or sorting application as might be found in the factory. The other approach is to build parallel-processing machines that divide the work so that many parts of the problem can be worked on at the same time. There are two popular methods of doing this. One is to divide the work into groups and let each group be handled

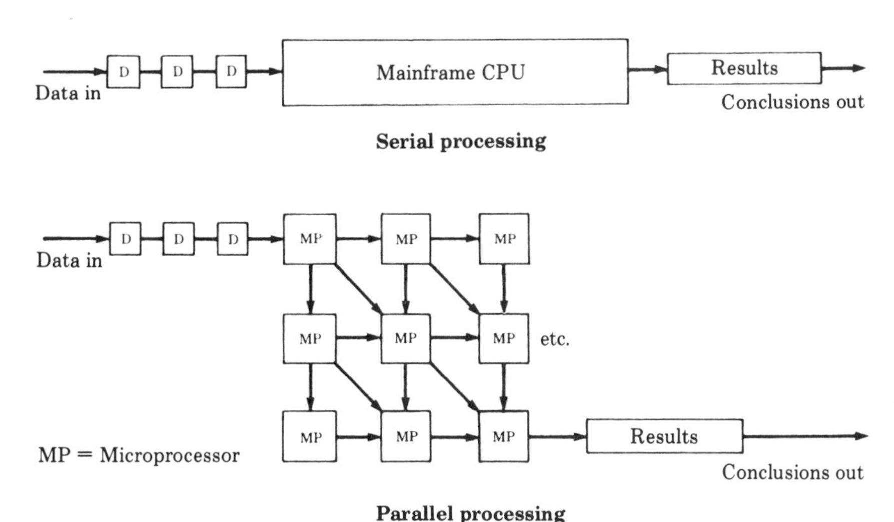

▶ ***Figure 9.13*** Serial versus Parallel Processing

by its own processor. The other method is to present all the work to a matrix of processors and let the work flow through the matrix, with each individual processor nibbling away at whatever work is left. This process is sort of like passing a cheese ball at a cocktail party. Machines of the latter type go by the name of ultra cube and hyper processors. A great deal of the current study on parallel processing is being funded by the government through aerospace and the military. However, many believe it will be cheaper to make machines with many individual, inexpensive, single-purpose processors than to try and build complex processors. When this happens, parallel processors may become almost as commonplace as the PC is now. Cheap processors will cause applications to explode. Figure 9.13 shows models of serial and parallel processing using the matrix approach.

9.5 Distributed Systems

Data Need to Be Distributed

When we talk about computer integrated manufacturing, we are saying we have set up an infrastructure that allows access and transmission of data from one PD to another. We are not saying that all manufacturing can be controlled by a single computer. This was tried early in the game and simply did not work. The central data base could never be made large enough and the necessary manipulations simply were not fast enough. As a result, a hierarchical data structure has evolved wherein local data are stored locally but are made available to whomever needs them, and the data are accessed through networks. Two aspects of the data are important to recognize. The first has to do with the quantity of data and the other with the quality of data. If it has the proper sensors, a machine is capable of generating data. These data may be information on position, velocity, and other physical considerations or the data may be statistical, such as part counting. For example, to function, a machine must receive instructions to turn on, begin cutting, stop when through, stop if something goes wrong, and so on. We have a situation where data can be generated, but some of the data may be relevant at only a given time and at a given location. Furthermore, we sometimes gather a great deal of data and manipulate them to come up with one answer. A host computer may want to know only the mean and standard deviation, which can be calculated at a local level. The data points themselves do not need to be passed on to the host. Since local level storage is usually at a premium, the data points can be periodically offloaded to bulk storage, such as a floppy disk, for archival retrieval.

▶ ***The Engelberger Model.*** In a typical robot application, we might find five levels of control. The Engelberger model, shown in Figure 9.14, lists these controls as system, work station, elemental move, primitive function, and servo. Servo, being the lowest, controls the position of the arm, for example. Primitive function has sufficient memory to be able to generate trajectories and the like. Output from the primitive function level is used as input to the actual servo motors at the lower level. Similarly, the output from the elemental move control goes to the primitive function control, and so forth. Each level then provides control commands to the level beneath it. Feedback data and information are generated at the lower level and passed on to the upper level to ensure that the commands were carried out properly. To provide feedback and to effect the control functions implies that some sort of data base is being utilized at each level of the hierarchy. With proper programming and networking, there is no reason why these data banks could not be made available to anyone within the system.

▶ ***Data Loggers Can Be Used.*** To be effective, the manager must know what data are available and when. Thus the manager must know in advance what is required and how to get it. As always, success is more likely if careful planning has been done. If many data points are needed, it might pay to add a data logger and data processor to provide the condensed information required at a higher level. Some advanced programmable controllers have provision for data logging, but most don't. How many parts were made and what were the means and variances of

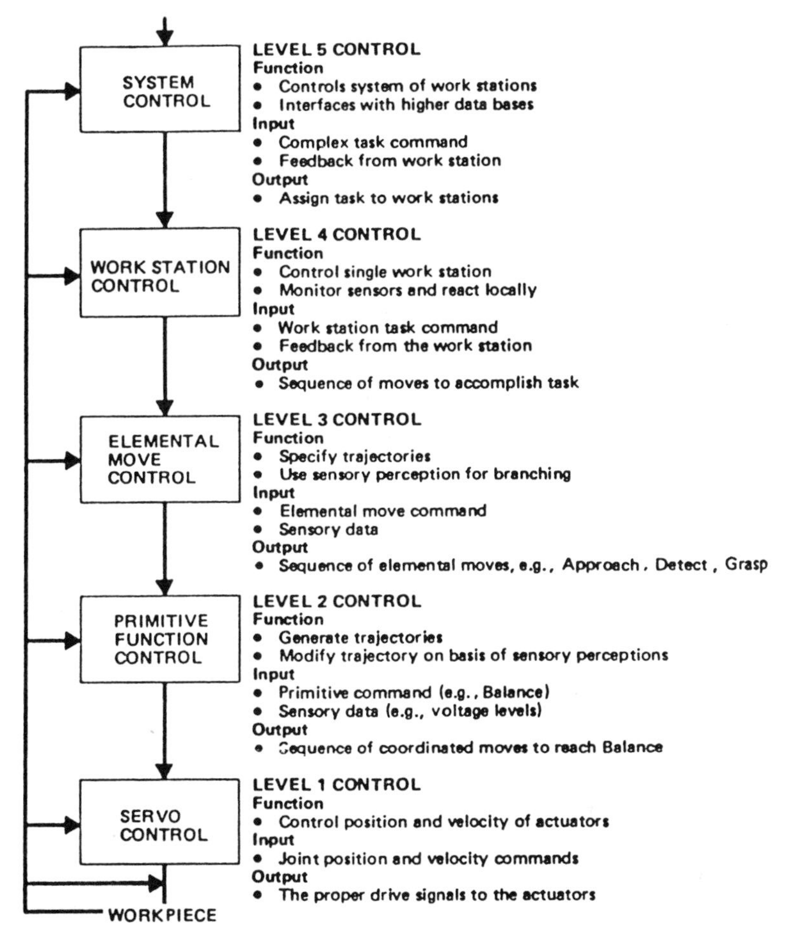

Shown is a hierarchical control system for robot installation.

► ***Figure 9.14*** Engelberger Model (Engelberger, *Robotics in Practice*)

the tolerances of the parts produced might be needed at a higher level to determine if a correction in the process is required. The data can be made available on a real-time basis for direct machine control or can be accumulated and stored for later retrieval.

9.6 Dictionary or Directory

► ***Data Bases Must Be Found.*** Data that are collected but that no one knows anything about are not very useful. Since data are collected

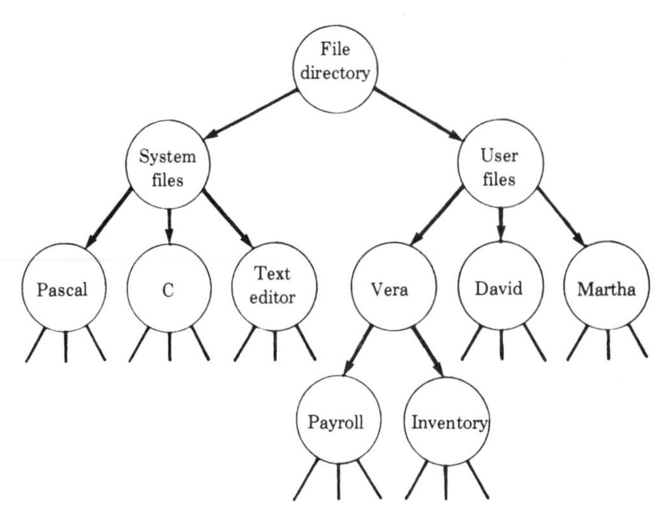

► ***Figure 9.15*** UNIX-like File Directory System (Singh and Naps, *Introduction to Data Structures*)

at various locations, it is a good idea to be able to work out a system so that the data pool can be accessed. This comes under the heading of a dictionary or a directory.

A word dictionary gives the definition of a word when you look it up. Here every word is organized in an alphabetical fashion. Some word dictionaries are organized in other ways, verbs and nouns being common. *Roget's Thesaurus* gives synonyms. In the same sense, we need to organize data pools according to some method and to tie the whole together with some sort of look-up table or other scheme. A common method, which is how UNIX is laid out, is to use labels that can be added together to form a pathway to the individual pool or data base sought.

A UNIX-like file directory system is shown in Figure 9.15. File Directory/User Files/Vera/Payroll will get us payroll information. Likewise, Louisville/milling/no15/count might be a sequence that gets us into the data-base network and directs our inquiry as to the number of parts produced since last asked by machine number 15 located in the milling department in the Louisville plant.

9.7 Distributed Data and Networks

► ***Data Are Usually Close to Where They Are Frequently Used.*** The concept of the data base in actuality is a series of small, local data bases scattered about the environment, each being accessible, leads to the need to establish some sort of hierarchical organization and a means to access the various data bases. We have mentioned the hierarchy associated with the control of a machine, and in the same sort of way there needs to be a hierarchy relating to the data bases themselves. The question of access by means of a directory and routing has to do more with the idea of getting into a particular data base on an occasional basis rather than utilizing that data base on a routine fashion. A station that is using the data base in a continuing and routine manner should, ideally, have that data base in close proximity and should own or control the data base. The logical place for the design information of parts should be in the CAD area. If someone in purchasing needs a drawing for a vendor, the data base should be accessible, but what goes into the data base and how it is formatted should be the responsibility of the CAD group. Similarly, the CAD base should be concerned with tool paths, but the actual instructions for cutting need to be machine specific and located near the miller. CAD personnel may interrogate the miller data base, but the care and feeding of the data base belongs to the miller.

► ***Need for Communications Networks.*** If we have a great number of data bases as part of the overall data-base system and if each data-base module can be directly accessed by every other data base module, the amount of wiring to connect the data bases to each other will be horrendous. A system of communication between the data bases is essential. Since the phone company has already solved many of the problems associated with network communications, the makers of computer equipment have drawn on this technology to solve the problem. Unfortunately, every maker has a different solution. As long as the situation is strictly local, a department or a plant will usually select a vendor and incorporate that system. On the factory floor we find, among others, Gould's Modbus and Allen-Bradley's Data Highway, and in the office we find Wang's Wangnet and IBM's SNA net. Difficulties arise when someone on the Wang net wants to get into a data module attached to the Modbus. This problem is being addressed and we will talk about it in a later section.

9.8 Access and Security

Data Security

The question of who gets to dial which data module is strictly political. Access depends on how valuable the data are and who determines who is allowed to access the data. Some information is truly crucial to the operation of a firm, and if that information were to get into the hands of competition, it could be disastrous. Beyond the firm, there may be military or national security needs for access control as well. If we are worried only about our competitors, the problem becomes much easier to handle. The general tendency is to be too restrictive in the control of data access. The fact is that you can keep information from your competitor for about six months, with three being typical, no matter what you do. Information is shared at meetings, people quit, executives like to brag, and so on. Also, some information is not really as important or proprietary as some operational personnel believe. A certain automotive supplier never allowed visitors in its plant. They did not want to lose all their secrets. After a change in management, visitors were allowed into the plant and found some operations that were so primitive as to be almost laughable. This company had failed to rely on their vendor's knowledge for the latest in information because their vendors were unable to see what the needs really were. However, some information is truly proprietary and needs to be guarded. The challenge is to separate the two. A better reason to limit access is so that a disgruntled person cannot alter or destroy the data in the data base. And if there is any possibility of monetary theft, we need to control access for this reason as well.

► ***Passwords Can Help.*** The usual method for access control starts with the assignment of passwords. A dedicated individual can usually find a way around this given sufficient time. Most passwords are short, typically six letters, so that people can remember them. Although any character is usually allowed, most people use letters of the alphabet and simple standard combinations like the name of their father. This makes the breaking of the password that much simpler. "David" is a lot easier to crack than "*&^%$." Assuming the password is valid, the next step is to restrict the data base to a select list of passwords. Sometimes it might be necessary to have two passwords to gain access, yours and someone else's. This can be subverted, like anything else, if there is collusion. Another technique is to allow certain passwords to be used only at specific terminals, such as only the sales manager's password being valid in his or her office.

► ***Encryption Is a Further Deterrent.*** Encryption and scrambling can be used and are gaining in popularity. Not only must the thief or troublemaker know the key, but the encryption can be changed frequently enough or made complex enough so that breaking it becomes extremely difficult. Even though the miscreant can gain access to the data base, the output looks like so much garbage. The basic encryption model is given in Figure 9.16. Note both that the sender and receiver need a key. Sometimes they can use the same key, but in advanced systems there are many keys for different people.

► ***Two Useful Methods.*** Although there are many different types and methods of encryption, two advanced systems seem to be fairly useful and extremely hard to crack. One type employs microprocessor

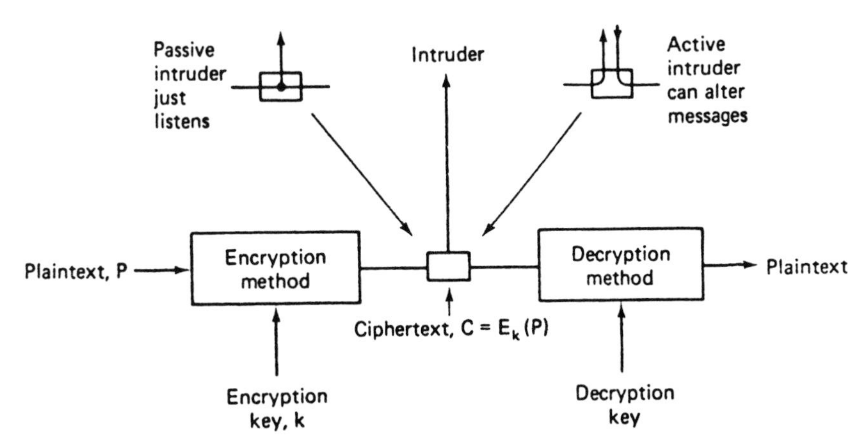

▶ ***Figure 9.16*** The Encryption Model (Tanenbaum, *Computer Networks*)

chips encoded so that bit paths are switched around as the signal passes through the maze. A similar module and identical program decodes the signal at the other end. Figure 9.17 shows this system as a series of P and S boxes. These can be switched around to suit the requirement. The other system uses factors of the products of extremely large prime numbers (a number that is not divisible by anything except itself and 1) and applies the products to transform the bits. The person at the other end is given a prime number that will allow him or her to decode the message. In theory, if you had a large enough computer, you could find the prime number that would crack the code, but by the time you did this the war would be lost. A popular method is to use pseudorandom numbers, which come as a standard feature in most computers. This will delay the hacker somewhat, but it is easily broken with a high-speed computer. It will, however, deter the mischief maker.

Undoubtedly, many other methods and means can be used to deter this type of interference. Since the idea is to provide security,

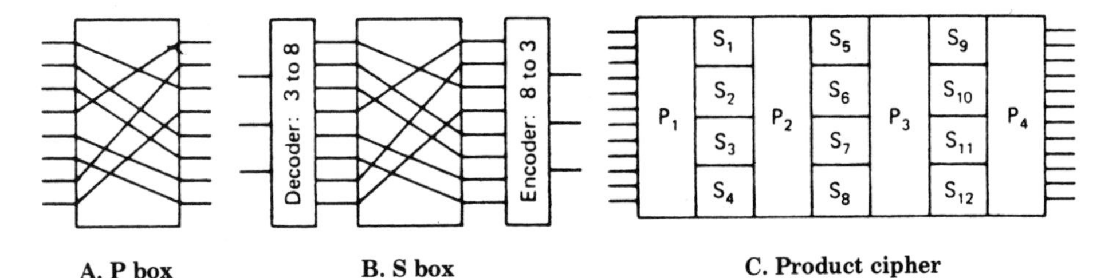

▶ ***Figure 9.17*** Basic Elements of Product Ciphers (Tanenbaum, *Computer Networks*)

anytime someone devises a technique to enhance security, this in itself becomes part of the security. Information is hard to come by. Even so, people are generally more resourceful than we give them credit for and our systems are not as secure as we think. It is best to assume that what you are trying to protect will be discovered sooner or later and, furthermore, someone is working on the challenge of getting into your system right now.

9.9 Protocol

► ***Protocol Tells How Communications Can Be Established.*** Protocol is a set of established rules that allows smooth and orderly transfer of information. Who sits next to whom at a summit conference is an example of where protocol needs to be established. In a manufacturing environment we need to establish a set of procedures and protocol so that the various machines can talk with one another in such fashion that all information gets sent and received, and important information gets sent and received before unimportant information. All this comes under the heading of computer network theory. Computer network theory is based to a large extent on the work and modeling done by Bell Laboratories and other leaders in the communications business. This work has been augmented by product development of the various suppliers of programmable devices. A great deal of history, tradition, and self-interest is associated with the development of network systems to transfer data. Because much of the work has been funded by various international governmental bodies, the self-interest feature includes a high order of national pride. Each country, working through its national standards organization, has developed standards that frequently do not agree with those in neighboring countries. In addition, leaders in the communications and computer business have established their own standards, some of which are in line with the standards established by their own national groups and some of which aren't. Nothing is compatible with anything else. To make a long story short, an event occurred in the fall of 1985 that may change all this.

Manufacturing Automation Protocol

In 1985, General Motors, who in the United States is farthest along with the integration of the factory, said, "Enough!" From then on, any supplier of equipment that has any kind of PD attached to it must conform to the Manufacturing Automation Protocol (MAP). GM had been working with some leaders in industry, such as Boeing, for about 5 years to establish what they considered to be the best of the hundreds of protocols available to serve as their guide. Some suppliers of PDs were more sensitive to the situation than others, but some who had a large vested interest in their own established systems were unhappy with the move. A large number of other user companies have been on the sidelines waiting to see what would happen. Now that a few have spoken, most of these other manufacturing firms cannot get on the bandwagon fast enough. The suppliers of PD equipment have also seen the light, and those who have not made announcements to the effect are working energetically to get their products to conform. IBM recently joined in and said they would go along with MAP despite the fact that their own SNA (Systems Network Architecture) network is widely used. It appears MAP is about to become the de facto standard in the industry.

▶ ***MAP Is Not for Everyone.*** Before we look at the nuts and bolts of MAP, we need to put in proper perspective who is using MAP. In the spring of 1986, there were several hundred members in the MAP users group. Of these, less than 100 are companies that are actual manufacturers and that can be classed as true users. The rest are suppliers. This has interesting implications, because there is a very large effort by the makers of equipment to foster the development of MAP. Implementation of a complex computer network is very expensive regardless of whose standards apply. This means, in reality, that only the very large Fortune 500 type companies will be viable prospects in the near term. However, because the products sold to everybody will be MAP compatible to accommodate GM and the other users, implementation of MAP into general manufacturing will take place gradually but with a high degree of certainty. This may tend to accelerate the implementation of CIM in American industry.

▶ ***MAP Uses the Token Bus.*** Let's look at how MAP works. Say we want to get a specific set of data from a computer in location A to the computer in location B. To do this, we have to add information to the actual data set itself so that the data will get to their destination. Think of it as similar to what happens when you write your friend. After you write the letter, you place it in an envelope and seal it. You then address the envelope and drop it into the letterbox. The post office picks up your letter and delivers it to its destination, where your friend picks up the letter, opens the envelope, discards the envelope, and reads the letter. From the viewpoint of the post office, the message in your letter is not important. What is important is the envelope. If you pay a premium price, you can get your letter delivered more quickly than if you don't. If you are a congressman, you can get your letter delivered for free. So we see there are rules or protocols established for the delivery of your letter.

MAP works along the same lines except it is electronic and all information is in digital form. Figure 9.18 shows the postal system compared to a typical electronic system. When the message is ready for transmission, it must contain a great deal of additional information

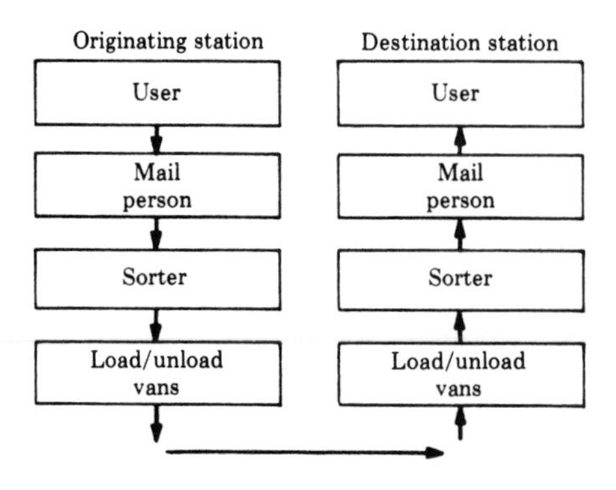

A. Postal system

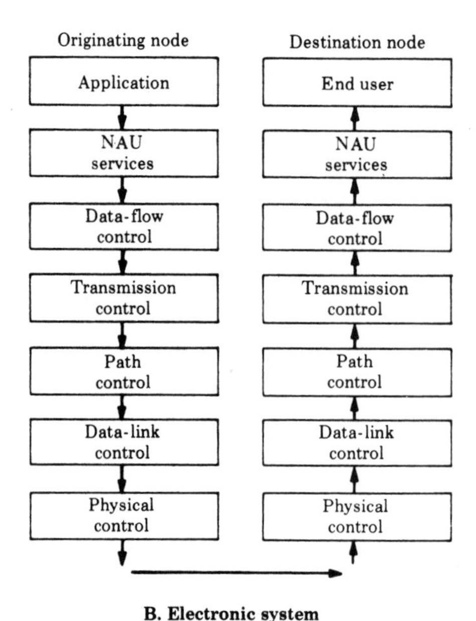

B. Electronic system

▶ ***Figure 9.18*** Electronic versus Postal System (Harish, *Understanding Data Communications*)

regarding who is sending it, who is to receive it, how long it is, and so forth. In the case of data transmission, we want to know if it actually was received at its destination. Provision to ensure that the message was received and the content was not garbled is included. We can have a receipt returned, an acknowledgment.

Figure 9.19 shows the message unit for a typical MAP message. Note that at least 20 octets (8 bits) of header data are sent before the actual message is sent. Because, unlike the post office, the transmission

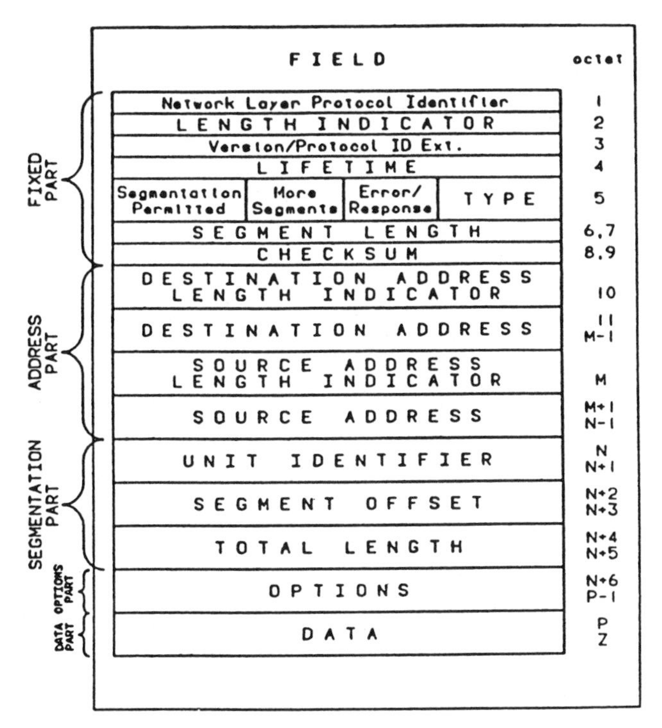

▶ ***Figure 9.19*** Data Headers (General Motors Corp., *MAP Specification Manual*)

media (the cable) can handle only one message at a time, provision must be made to determine who gets to send what and when. MAP handles this by using a token bus system. Let's see what this means.

▶ ***There Are Several Token System Topologies.*** There are many different kinds of configurations for networks. Possible topologies are shown in Figure 9.20. The loop or ring, the star, and the tree or bus are the most common. A ring is a circle where every computer is attached to the ring. Messages to computers other than an adjacent computer must still pass through the adjacent computer. A star has one computer in the center that redistributes the message to the proper party. In a bus system, all computers are hooked to a common line so that all computers can receive messages directed to all other computers. If two computers want to transmit at the same time, this can lead to problems.

▶ ***How the Token Bus Works.*** One way to handle the situation is to let each computer transmit whenever it feels like it as long as the line is

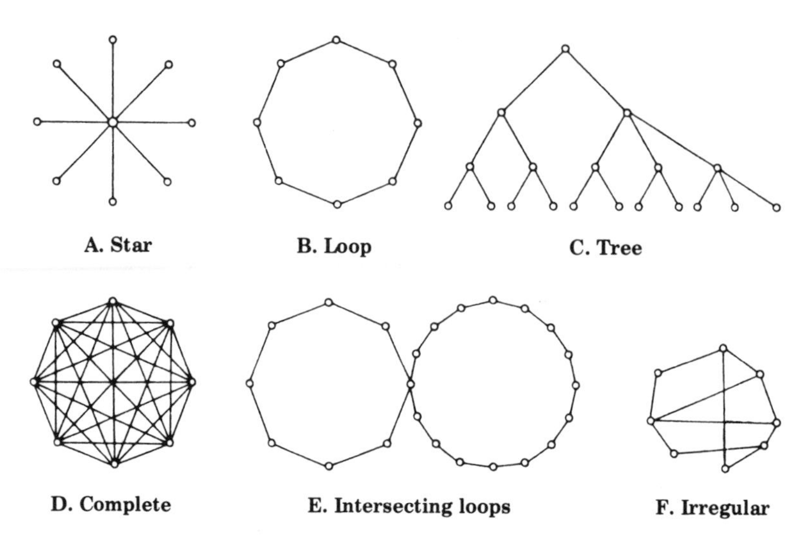

► ***Figure 9.20*** Topologies (Tanenbaum, *Computer Networks*)

free. However, two computers can still transmit at the same time, and so a collision may occur. After a delay, the computer will know the message did not get through because it did not receive an acknowledgment, so it retransmits. Up to a point, odds favor the next message will get through. This is called a contention system and is widely used. With MAP, a special character set or token is passed around the network, and if a particular computer has a message to send, that computer grabs the token. Think, "You can't ride the bus without a token." Although the physical configuration is a bus, logically the system is a ring. Figure 9.21 shows the physical layout for a small plant. The token is passed along the bus in a prescribed sequence until it is picked up by a computer that is allowed to grab it and that has a message to send. If the computer has no message, it passes the token to the next party. Allocation of the right to send is based on the need to send, which is dependent on the urgency. Notification of the impending collision of a robot arm with the floor needs to take precedence over routine statistical data.

The ISO/OSI Model Is the Heart of MAP

The MAP protocol is based on a model designed by the International Standards Organization (ISO). This ISO model is classified as an open systems interface (OSI) system. We can refer to it as the ISO/OSI model. This model has seven layers, starting with the physical layer and going up to the application layer. The eighth layer is the interface between the user's program and the application layer. At each layer as we move from

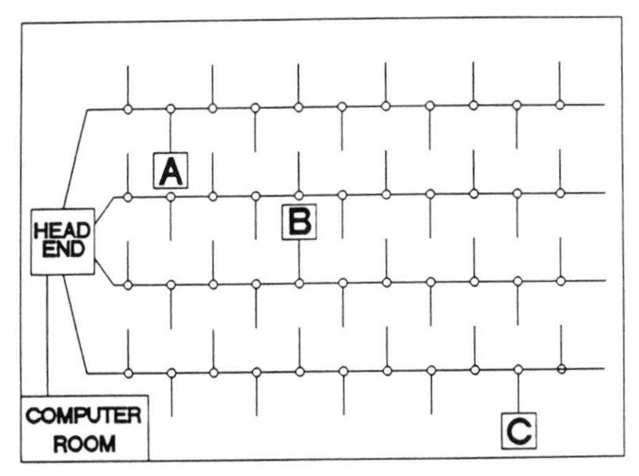

Shown is a physical layout for plants shorter than 1000 feet.

▶ ***Figure 9.21*** Network Physical Layout (General Motors Corp., *MAP Specification Manual)*

the user down to the physical link, we add more and more address and transmission information to the data packet or message that we actually type into our terminal or that our microprocessor puts out. We also need to do some signal conditioning so that the network can accept the complete package. We put address, synchronization signals, routing information, and error-handling routines onto the message. At the physical layer, we encode the message and put it into acceptable electrical form. The message is eventually sent and received at the other end, where the process is reversed. All the address and other information is stripped as the message moves up through the seven layers until what is displayed on the receiving screen is only the message as entered at the sending terminal. All the intermediate information attached to the message is transparent to both the sender and the receiver. Figure 9.22 shows how the seven layers are nested and applied.

▶ ***ISO/OSI Is Still Being Designed.*** The ISO/OSI model is intended to be the framework upon which the specific specifications hang. The model reflects all the steps that need to be taken to move the message from one point to the other. However, the detail of what is actually included has not been finally determined. At this time, only the first two or three levels have firm written specifications. On the lowest level, or physical layer, the protocol that has been established is IEEE 802.4, which is a broadband cable configuration. This cable has a 300-megahertz bandwidth and can handle transmission rates of 5 Mbits per second. As stated, the access method is by the token medium. At the data-link level, the next higher level, IEEE 802.2 has been selected as

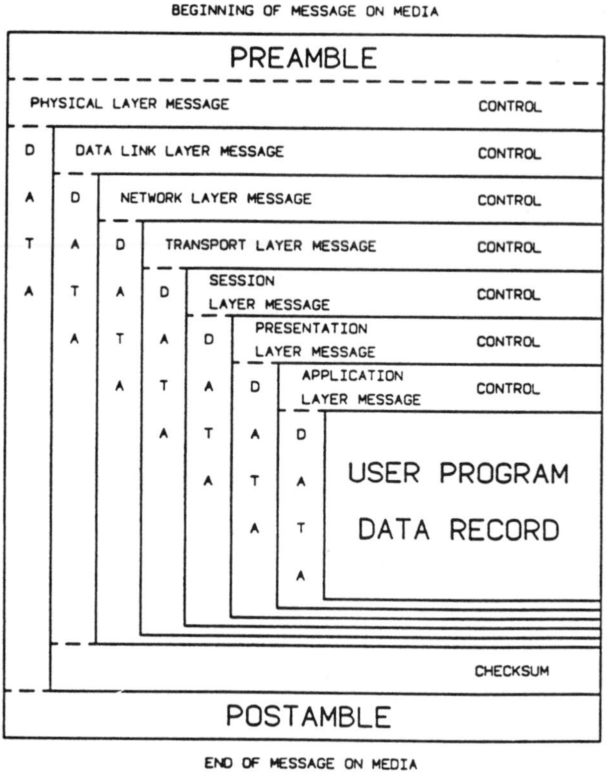

BEGINNING OF MESSAGE ON MEDIA

PREAMBLE

PHYSICAL LAYER MESSAGE CONTROL

DATA LINK LAYER MESSAGE CONTROL

NETWORK LAYER MESSAGE CONTROL

TRANSPORT LAYER MESSAGE CONTROL

SESSION LAYER MESSAGE CONTROL

PRESENTATION LAYER MESSAGE CONTROL

APPLICATION LAYER MESSAGE CONTROL

DATA DATA DATA DATA DATA DATA DATA

USER PROGRAM DATA RECORD

CHECKSUM

POSTAMBLE

END OF MESSAGE ON MEDIA

▶ ***Figure 9.22*** Nesting of Layer Protocols (General Motors Corp., *MAP Specification Manual*)

the protocol; it deals with how data are to be handled to reduce error rates. The other levels of the ISO/OSI are in various stages of being standardized. Keep in mind that industry is not going to wait until the standards are set. Each plant will have established a local network that is run using the standards established by the manufacturer of the network equipment or the plant management of the company. In the future, equipment will have to be retrofitted to conform with the standards as they are implemented, using applicable hardware and software interfaces. Smaller companies that have few plants or companies that have many divisions may never move up to a MAP protocol. They will likely continue to use whatever they already have in their proprietary local net.

Gateways Get Us between Networks

If we want to move from one network to another, we must provide a gateway, bridge, or router to get from device A to device B (Figure 9.23).

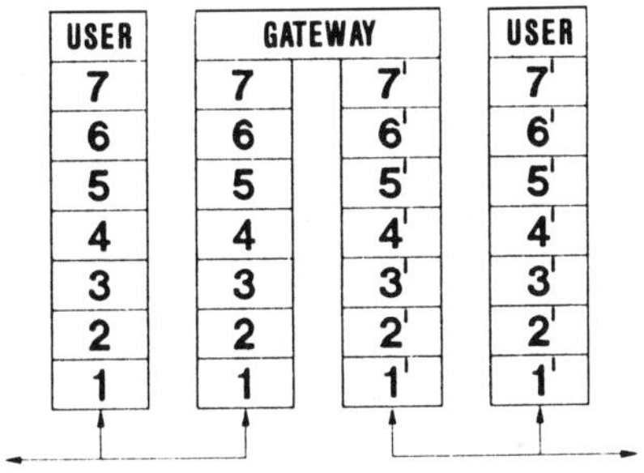

USER	GATEWAY		USER
7	7	7'	7'
6	6	6'	6'
5	5	5'	5'
4	4	4'	4'
3	3	3'	3'
2	2	2'	2'
1	1	1'	1'

Network A **Network B**

▶ ***Figure 9.23*** Gateway Architecture (General Motors Corp., *MAP Specification Manual*)

A gateway would be used if there were two distinct and separate networks that needed to talk with one another through radio transmission or long-line telephones. The sender would add transmission information to get onto the MAP network and would send the packet to the gateway, a device suited for getting the message into a form the radio transmitter can handle. At the other end, a similar gateway device accepts the message and puts the MAP information back onto the message so that it can be handled by the local MAP network. After the message is received by destination B, the transmission information is stripped and the message is presented to the user.

Figure 9.23 shows a gateway network interface. Bridge and router interfaces, which are used to communicate between local systems, do not need to go through all seven layers and are simpler to effect. Figure 9.24 shows how a large company system might be configured using the three interconnecting links.

9.10 Summary

It is important to have the right kind of data storage because without a cogent and uniform scheme the computer will not be able to use data from other sources. It is also important to put correct data into storage and to not load up the memories with unnecessary information.

Data-base management relates the various different elements of data so that the data can be accessed, manipulated, and presented to the user. There are many different kinds of data bases, but the two most common are relational and hierarchical. Each has its advantages and disadvantages, but the main thing is to be able to pick out the data needed from all the data in the system. Now data are distributed, which

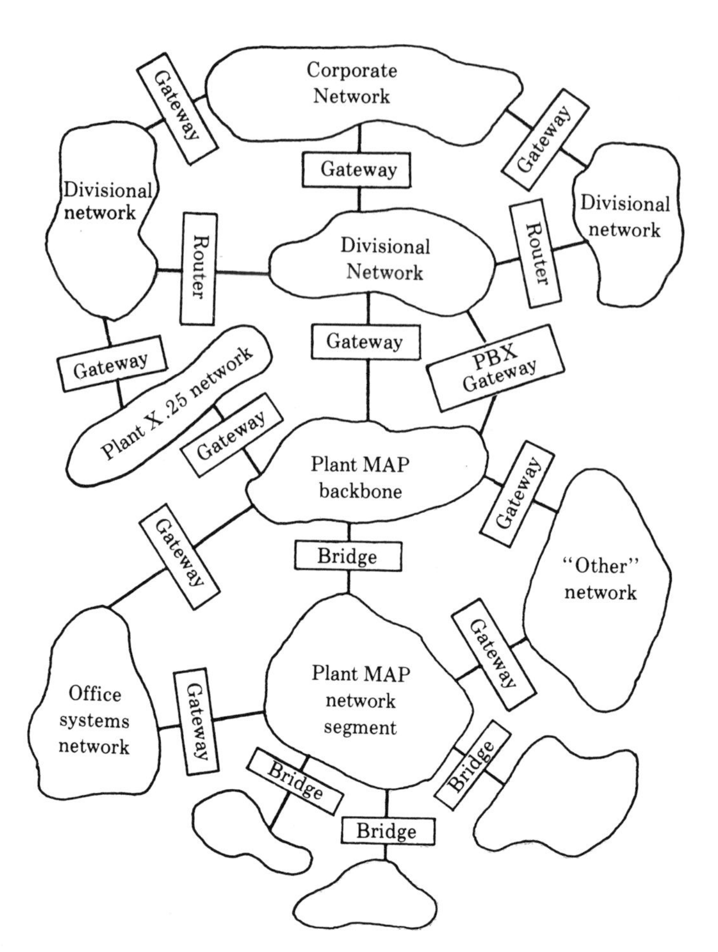

▶ ***Figure 9.24*** An Inter-Facility Network (General Motors Corp., *MAP Specification Manual*)

means that they may be in a remote location from the computer that needs them. Access is very important.

Artificial intelligence data bases are extremely large. Because it takes an appreciable amount of time to access data and to use programs, all of which are done in a sequential fashion, conventional computers are too slow for real-time AI use. One solution is to use parallel processors, which can handle many calculations at the same time.

To access data in a network, we need to develop a protocol. This establishes a set of prearranged rules of communication. Many protocols have been developed by various manufacturers, but the one most likely to be used is MAP. This system uses a token bus and is designed around the ISO/OSI model.

9.11 Exercises

1. Why do we need to be concerned about the data in the data base?
2. What does data-base management encompass?
3. What does GIGO mean?
4. What does a local area net do?
5. What is the de facto network standard in most small companies?
6. Name some applications for sequential files and random files.
7. Describe a typical relational data base and a hierarchical data base.
8. What is the central problem in AI?
9. Where is AI used in the CIM factory today?
10. What is the difference between an expert system and AI?
11. What are the two general parallel-processing techniques?
12. What is the difference between a central computer data base and a distributed data base?
13. Describe the Engleberger model of hierarchy.
14. What does a data logger do?
15. What is meant by dictionary or directory?
16. Where should a particular data base be located?
17. Why don't we just connect every machine and data base to every other machine and data base?
18. Why are there so many different kinds of networks?
19. What are some of the people problems associated with security?
20. Name three reasons why you would want to have a secured data base.
21. What are two generally used encryption methods?
22. Regarding security, what is the best thing to assume?
23. What does protocol mean and how is protocol applied in the factory?
24. What does MAP mean?
25. Who was the leader in implementing MAP?
26. Who is using MAP right now?
27. What is the key technology with MAP?
28. What kind of information is attached to the message?

29. Name several kinds of token system topology.

30. What is the contention system?

31. How does the token system work?

32. What does ISO/OSI mean and how many levels are in this model?

33. What does a gateway, bridge, and router do?

10

The Role of Executive Leadership

10.1 Introduction

Emphasis in earlier chapters has been directed toward the goal of implementing CIM through hardware installations. We have looked at the various mechanical and electrical components that go into an automated factory. We looked at the robot in depth as a model of the kinds of machines and material-handling equipment we need. We covered some of the communication requirements and data structures that are essential for a CIM facility. In short, we have discussed all the technological aspects that go into a CIM facility. Now we need to turn our attention to the management aspects of a CIM facility.

To get a clearer picture, much of the discussion that follows compares the worst of the traditional U.S. factory methods with the very best of the Japanese methods. Many U.S. companies do employ the latest advanced methods and many Japanese companies are still in the managerial samurai age. The true situation is that neither black nor white but various shades of gray occur in both countries. However, lest we get too complacent and congratulate ourselves on how far we have come, it needs to be pointed out that Japanese productivity is still higher, their unemployment is lower, their people have better health and social services, their currency is stronger, their GNP growth rate faster, and their balance of payments better than that of the United States. Clearly, they make an attractive model for us to examine.

10.2 Commitment

► ***Success Comes from Planning and Commitment.*** There are two hallmarks of the Japanese method of increasing productivity in the factory. The first is planning and the second is commitment. Before the production line is built, it is essential that all the homework be done. A thorough analysis of the market will have been made to see what the customer wants and how much he or she will pay for it. Answers to questions about models, style, color, benefits, and features will all have been resolved before the final design has been determined. If test marketing with prototype units is required, these units will be fabricated in pilot operations. Concurrent with the design of the product will have been all the process design necessary to fabricate the product. In short, a great deal of in-depth planning will have been done before the product gets onto the line.

► ***The Japanese Get All the Data Before They Proceed.*** Executive style is vastly different in Japan than in the United States. A top manager will ask an oblique question of his subordinate such as, "Do you think we could market such and such in the United States?" What he is really saying is, "I believe there is a market for such and such and I not only want you to find out all about the market, but I also want you to figure out how to make the product, what it will cost, and so on, and give me an in-depth presentation on the subject." The subordinate, through questioning, gets every scrap of information he can and only then presents it to his superior. This process takes many months, but when the presentation is ready it is accurate and thorough.

A number of years ago the industrial chemical division of a large consumer products company had a minor product that they were trying to license to the Japanese. The product was a "dog." The representatives of the Japanese trading company arrived and wanted to know all about the product and the markets. The division managers tried to present the positive aspects of the product and gloss over any deficiencies. Despite their best efforts, the Japanese were able to discover all the bad features. Each member of the team asked people at all levels in the company certain specific questions. Rather than go out at night, the visitors met in a motel room to compare answers. They were looking for inconsistent answers. The next day they made it a point to have another member of the team ask the same question, phrased differently, and they switched team members around with the interviewees. The U.S. division managers caught on to what was happening after about three days but by then it was too late. One could tell from the kinds of questions the Japanese were asking that they had done a thorough market research before they arrived.

▶ ***Put in Enough Resources to Get the Job Done.*** Once the Japanese decide to go after a market, they put in whatever resources are necessary to provide applications information. If we decide to market a resin for paints, for example, we will assign three or four chemists to develop illustrative formulations showing how this product can be used. The Japanese will assign 75 to 100 formulators on the same problem. We may have the same number of formulators, but ours will be working on many different products and for many different end uses. The Japanese use a rifle to target one market, and we use a shotgun to target all possible markets.

▶ ***Specifications Come from the Customer.*** After we determine what we are going to sell, to whom, and for how much, we should give consideration to the question of product design and manufacture. If we have done a good job of market research, we will, at the same time, be able to write a set of performance specifications. These product specifications should have been set as a result of extensive conferences with the customer to determine exactly what they require. This approach is a radical departure from the usual method of setting specifications. Many companies set specifications on the basis of what they think the customer needs or, more frequently, what they can make. Products are designed, tested, and brought to the market in many instances without ever finding out whether there is a real need for them. A good example is the Edsel automobile.

▶ ***Work Out the Process Before the Factory Is Built.*** Many companies engage in the practice of doing product and process prototype design work while they are in the process of building the full-scale factory. Sometimes it is necessary to develop a prototype model in order to test it. The prototype model design and building should be separate and distinct from the production and process design work that takes place. The prototype design work should be done first, followed by production and process design. Figure 10.1 shows the sequence of idealized product development flow. This sequence is particularly relevant if the product is a new product instead of an improvement of an existing product design. The engineering and other technical information learned in the prototype and test stages is used in the following product and process design stage. The idea of finding out what the customer wants, translating these ideas into specific specifications, and building a prototype model and testing it, both in the marketplace for customer approval and in the lab for durability and other product characteristics, before starting on process and production engineering seems so fundamental as to be obvious. However, it is a fact that U.S. industry does not use this approach very often. There are some notable

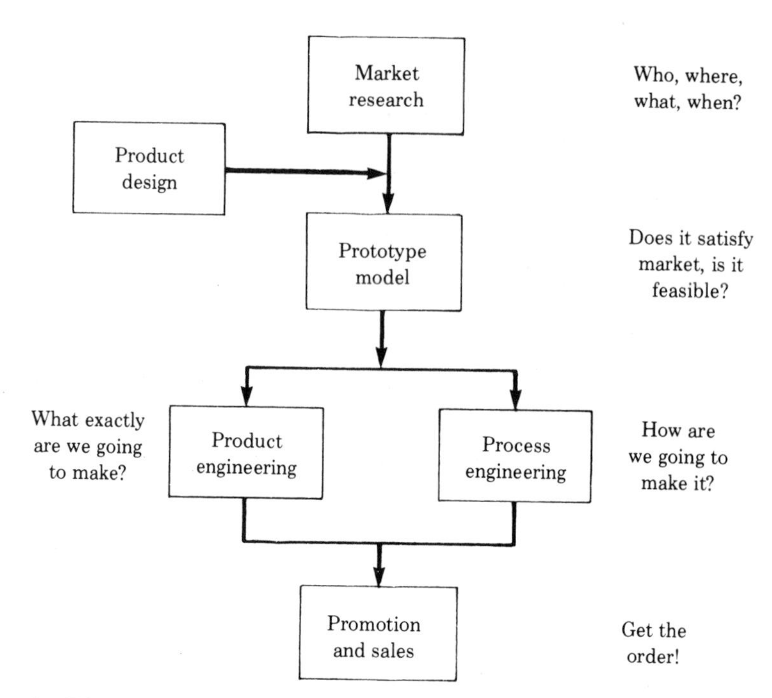

▶ ***Figure 10.1*** Idealized Product Development

exceptions, of course. Certain U.S. companies such as Hewlett-Packard, IBM, 3M, and many others do an outstanding job of planning their operations and managing their employees. These companies have also had many years of consistently high profits and have remained globally competitive. On a 5-year project, the Japanese will typically spend 4 years in the planning stage of market and product research, product and process design and engineering, and other front-end tasks, and will spend 1 year building the factory and eliminating unforeseen production bugs. We spend 1 year on the front end and 4 years getting the factory to produce good parts. The Japanese method is less costly.

▶ ***Culture Plays a Large Part.*** Lest we assume that our managers are less smart than their Japanese counterparts, we need to consider some cultural differences. Our MBA (masters of business administration) schools stress the idea of maximizing short-term profits. Until recently, little consideration was given to the idea that we need to balance short-term results with growth over the long range. The U.S. managers' pay schedule is predicated on how well they do right now. Since our managers are interested in promotion, which means higher pay, most individuals will expend their energy getting promotions. This is referred to as the "fast track." One way to get on the fast track is to generate excellent financial results immediately. Since excellent results in one department or area may be had at the expense of other departments or areas, this approach, although good for the individual, may not be good for the whole company. The term for this idea is suboptimization. For example, one way to cut costs that will enhance profits is to fire staff people. This popular method will enhance short-term profits but will result in a degradation of the overall operating efficiency over time because services that need to be provided for future welfare won't be provided. The manager who uses this approach is gambling on the premise that he or she will be promoted out of the department or will get a much higher paying job at a rival firm. Financial security and independence in the United States come from position and professional reputation. Many a U.S. manager will not hesitate to cut ties with a present employer to get a better job somewhere else. This does not promote much group loyalty or company loyalty. Furthermore, the U.S. manager knows the company will not hesitate to fire him or her if business deteriorates, if a take-over occurs, or for simple office politics or many other reasons. Therefore, he or she is obliged to look out for himself/herself and try to achieve financial independence and security. Top executives protect themselves with lucrative terminal contracts called "golden parachutes."

The whole situation in Japan is different. The manager is hired for life in larger companies. The cultural emphasis in all activities (schools, sports, military, business, etc.) is on group welfare and participation. The Japanese manager does not have the pressures to produce for the short term only. Pay and promotion are more dependent on how well the manager gets along with the group and how well the group is motivated to do their best. In this environment, the manager can devote more time to optimizing for the good of the company and less time suboptimizing for his or her own personal glory. Figure 10.2 compares management systems.

Japan	***United States***
Teamwork	Star system
Group goals	Individual goals
Optimization	Suboptimization
Long-term outlook	Short-term outlook
Company loyalties	Professional loyalties
Lifetime employment	Tenuous employment
Company unions	Trade unions
Education and training	Carrot and stick

► ***Figure 10.2*** Japanese versus United States Management

The contrasts presented are intended to show the two extremes of the situation. Again, the situation in the average U.S. company is not as bleak as presented nor is the situation in the Japanese company as favorable as described. Many U.S. companies are starting to recognize the need for incorporating the long-range approach and are beginning to pay managers on the basis of overall long-term performance. Also, trade articles point out that many individual Japanese managers would like to break out of the group and become stars on their own.

▶ ***JIT/CIM Demands Cooperation and Commitment.*** This leads us to the question of management commitment. Since JIT and CIM require extremely close cooperation between individuals, groups, and departments, factory management in particular will have to make radical changes in how it operates. These changes cannot come about until and unless top management recognizes the need to change and is willing to make the commitments to do this. Top management needs to "go to school" in the sense that to practice CIM and JIT programs they need to understand what the programs are all about. Top management knows that CIM and JIT can provide benefits because they see what is happening to them in the international marketplace. If you ask top management people if they believe in JIT and CIM tenets, they will undoubtedly answer yes. It is, however, a long way from saying that you believe in JIT and CIM and taking the appropriate steps to implement this philosophy.

▶ ***JIT/CIM Doesn't Happen Overnight.*** The transition from JIC to JIT is both long and very expensive. Since top management is judged frequently in public corporations by the performance of the stock in the stock market and since stock prices reflect short-term conditions, the same pressures to suboptimize exist in the executive suite as they do farther down the line. To get over this hurdle, the chief executive must commit the appropriate financial and human resources. In so doing, he or she must cope with the personal trauma of altering his or her power base. To implement CIM and JIT, it is likely that a total overhaul and alteration of the corporation need to take place.

▶ ***Participative Management Is the Key.*** A great deal of participative management must be made possible. Authority must be delegated farther and farther down the line, and chief executives must be willing to support decisions made by subordinates with which they do not agree. Many people, especially those who are used to a paternalistic and dictatorial mode of operation, may find it hard to make the transition. Their attitude toward enlightened subordinates is "convince me, if you can." In extreme situations, they may never be able to make the change and must be replaced. Unfortunately, this does not happen until the situation becomes catastrophic. Excessive labor strife is sometimes a symptom, with protracted strikes, high turnover, absenteeism, and poor productivity the overt signs of trouble. What happened to a steel company recently is a good example. The situation became so intolerable that problems were aired in the *Wall Street Journal* and the chief executive officer (CEO) was removed. The situation is not always that clear or certain, however.

Company stockholders, the same people who are looking for large and quick profits, need to elect board members who will take decisive action to ensure that top executives do not suboptimize. The paradox should be clear in that the people who will benefit most by having CIM and JIT implemented will most likely be the ones who will give the strongest resistance to their implementation, particularly since there will be large negative cash flows for long periods of time.

▶ ***Implementation Will Be Difficult.*** Assuming the CEO embraces the tenets with zeal, the problem of management commitment at the lower levels becomes no easier. Hundreds of years of tradition must be changed in order for the JIT and CIM philosophy to become a reality. Companies with competitive troubles will change more quickly than those that are doing well. Even so, studies have shown that the group most resistant to the introduction and implementation of robotics is middle management. Fear of loss of pay and status are the main problems. Lower and middle managers are paid on the basis of how well they perform in the short run. Because there have been so many technical problems implementing robotics and other advanced technologies into the factory, it is fairly clear that this approach has not provided the results that were anticipated. When things go wrong, who gets fired first? Certainly not upper-echelon executives, who will protect themselves, and certainly not the blue-collar worker, who is protected by the union. Even if the middle manager does not lose his or her job, if the program is not a success, the upward thrust of that individual's career may be halted. People are promoted in large corporations on the basis of the success of whatever projects they are involved in. If the risk is too large, thinking managers will demur and try to become associated with projects that are certain to succeed.

▶ ***The Middle Manager Is in a Tough Spot.*** An interesting article in the May 13, 1986, issue of the *Wall Street Journal* talks about GM factories that were designed to produce 60 cars per hour producing only 30. Similar conditions exist at other automotive companies, as well as at General Motors. The point is that the implementation of the technology is much more difficult than it first appears. When the CEO of your company asks you to head up a program to implement a CIM factory and he has been led to believe it will take six months and cost X, are you going to tell him it will take 2 years and cost 4X? This puts the middle manager in a difficult position.

▶ ***Middle Managers Build Empires.*** A manager's pay is largely determined by how many people report to him or her either directly or through subordinates. Thus there are tremendous pressures to build empires. The more people you have working for you, the higher your pay and status are. In fact, it could be argued that the first thing you should do when you get a promotion is to try to create slots beneath you. Your boss might go along with this because this puts another level under him, as well. The extreme example is the military, which has grown at the officer level exponentially. However, the lure of CIM is that it will eliminate labor and will let the company produce better quality goods. Attention has been on the blue-collar worker, but it is the clerical and skilled functionary that is most vulnerable. Consider what has happened to the detail draftsperson with the advent of CAD. People who are skilled functionaries, such as purchasing expediters, warehouse personnel, and inventory control managers, are especially vulnerable. As these people are displaced by machines and computers, the managers of these people will perceive their power base is being eroded. They recognize that machine tenders do not get paid as much or enjoy as many privileges as do people who manage other people. Is it any wonder then why there is so much resistance to change at the middle-management level?

On the other hand, entry-level managers coming out of our technical schools are in a strong position. These people will be responsible for implementing CIM and JIT in the factories. Those with vision will become expert in the field and their jobs will be secure. Also, being technical specialists, they will earn above-average pay. One discouraging note is that promotion may come more slowly as more people are entering these jobs now, business could be better, and levels of management are being compressed. The stated goal of the CEO of Dana Corporation, for example, is no more than five levels of management. Some companies still have as many as 15 levels and they will have to change. If a middle manager is effective, he or she will ensure that the company stays profitable and competitive. The corollary is that if a company is not profitable and competitive it is the middle manager who will disappear, which is exactly what is happening. By becoming proficient in the execution of modern methods, the low-level manager will be able to keep his or her job when promoted to middle management.

10.3 Overdesign

▶ ***Americans Are Not Waste Conscious.*** As a nation, we tend to overdesign. We frequently look for complicated solutions to simple situations. We see this not only in product and process design but in the overall plan of a factory itself. Our nation has never experienced any real shortages and as a result we waste a great deal of material and other resources. Until fairly recently, the idea of recycling was not very popular. All this has changed because we are now competing in global markets, and to compete with foreign countries and sell in foreign countries we need to make use of all our resources and not waste anything. Since our culture is not geared to saving, we naturally tend to be wasteful. It is curious that we are now urging the Japanese to save less and waste more. Perhaps, if we can convince them to downgrade their efficiency, our relative position will improve. Even so, we need to think of ways to better use the resources we have.

▶ ***The Japanese Can Beat Us in the United States.*** One argument for Japanese success is that they have a smaller country, they import all their materials, and there are very real differences in how they do things because of the particular environment and culture. However, in the same article cited earlier, we discover that Mazda put in a plant in Detroit that produces about the same number of cars as the GM Hamtramck plant, but the Mazda plant cost 25% less to build and employs 70% of the workers to do the same job. The plant cost less to build because the Japanese used less complicated technology. It employs fewer workers because the workers will be used differently than the workers at the GM plant. The bricks come from the same brickyard and the workers belong to the same union. No matter how you look at it, we are being beaten by design.

▶ ***Quality Is Conformance to Specifications.*** We have discussed the idea that the customer sets the specifications and it is up to us, as the manufacturer, to come up with designs that meet the specifications. If we agree that the definition of quality is conformance to specifications, then we must try to figure out how to design the product so that it meets the specifications. At the same time, we need to develop the process to make the product. An interchange must take place between the product design people and the process design people from conception all the way to completion. As a part of process design, we need to consider both the machine aspects and the people aspects of the process. The Japanese still use people in their factories and they only automate where it is profitable to automate. The reason our CIM factories don't work as intended is because the machines, which are tightly coupled to one another and are much too sophisticated for the task, were installed before they were proved in practice. An example might be a vision system that lacks adequate depth perception, resulting in welds that don't get made because the head missed the target. Again, it comes down to detailed planning before the machines get out onto the factory floor. A key part of the planning is to ensure that we use only a minimum level of complication. In an American-designed plant we might find a massive computer hookup to handle the Kanban system (assuming we installed a Kanban system in the first place), but a similar plant somewhere else might use a piece of plastic pipe and colored ping-pong balls to convey the same messages. It is true that the Japanese use many more robots than we do, but their robots are very much simpler. They have truly learned the meaning of KISS, or keep it simple, stupid.

10.4 Looking at CIM as a Cure-all

► ***We Need to Optimize.*** The question then centers on the idea of optimization. Optimization is the bringing together of all the elements in a system to either obtain the greatest benefits or to obtain the lowest costs. It is possible to achieve an individual cost reduction in one area but at the expense of other areas, resulting in the total costs being higher than they should have been. When this happens, it is suboptimization. The GM/Mazda situation illustrates that GM suboptimized the Hamtramck operation.

► ***CIM Is a Two-Edged Sword.*** Undoubtedly, GM was sold on the idea that CIM was going to solve all their competitive problems. This may well be the case in the future once all the problems have been solved to the point where the systems can be easily implemented. However, in the spring of 1986, CIM was not the cure-all. GM employs some of the brightest and best people in the country, who are willing to take risk, with the risk being open to public scrutiny. Their vision is that implementation of CIM is a certainty and they admit that they may falter along the way. Part of the situation may be because GM, through its GMF and other CIM equipment manufacturing affiliations, is willing to accelerate the process. A factor cited in the *Wall Street Journal* article for making these commitments is that the auto companies have had some very profitable years and can afford learning ventures of this kind, whereas they may not be able to do so in the future. In any event, it looks as though the Japanese have been able to think of CIM as just another tool at their disposal and to use those parts of CIM that can be easily applied. Their approach to the situation is more holistic than ours because we tend to think of CIM as an end in itself. They are better able to see the big picture.

10.5 Controlling the Right Things

► ***Ask the Right Questions to Get the Right Answers.*** We spoke of Pareto earlier. One recurrent theme is to concentrate on the big things and let the smaller ones take care of themselves. If top management tells the troops that the plant is to be automated with CIM equipment, then most assuredly the plant will be automated with CIM. However, someone must ask at the onset if that is going to solve the problem. This means that whoever is asked must not only be able to give the right answers but must also be free and willing to give the right answers. There is no one person or group who can do this. The CIM vendor will tell you his equipment will solve all the problems. The consultant will tell you his knowledge will solve all the problems. The problem is that all the answers will be different. The person who asks the question, presumably the CEO, will have to provide his or her own answer by piecing together bits of information from all sources. One function of the executive then is to gather information and to act on this information after analyzing it. Since there will be much more information available than can ever be used, the trick is to distinguish between information that is important and that which is not. The term for this ability is perspicacity, the art of acute mental vision.

All the shrewdness in the world will not help if the information is late, inaccurate, or insignificant. Toward this end, it is essential that sound control procedures be established to provide this information. Much of this information comes from the Management Information System group, frequently accounting and financial in origin. A debate is currently under way among accountants as to whether traditional cost information being supplied to the CEO is relevant. In manufacturing accounting, the cost of direct labor is used as the basis for the allocation of manufacturing overhead costs, such as utilities and maintenance. Overhead is then allocated on the basis of so many units of direct labor going into that product or operation. If the direct labor goes down, the unit overhead rate goes up. Also, part of the problem is that we define productivity as output per unit of labor. Since in a CIM plant there is very little labor, we come to the conclusion that the more CIM we install the higher our productivity (good) and the higher our overhead rate (bad). Last, but by no means least, is the problem that most capital decisions are based on some sort of return-on-investment criteria. But how do you quantify the return on such things as better quality, expanding our market share or remaining competitive, lower employee turnover, improved employee worth through training, and upgrading the level of technology? Not only do we need to ask the right questions, but we also need to ensure that we are getting useful answers.

10.6 Summary

There will be significant differences in management as we move into a CIM environment. Most important is the support and commitment required from the chief executive officer. Not only must enough time and money be made available, but the CEO must also be willing to give up some power and control to subordinates. Extensive planning must be done to ensure success.

Implementation of CIM will result in disruptions within the organization and different people will be affected in different ways. The group that is most vulnerable is middle management. These people are threatened because successful integration will allow computers to do many of the tasks now being done by this group. Lower-management people will be required to implement the changeover to CIM and will remain secure as long as they keep up their technical skills.

One effect of a CIM/JIT operation is that quality problems will be obvious. Quality means conformance to specifications, which the customer ultimately sets. The specifications, in turn, should reflect a level of utility that is neither too stringent or too lax. If management becomes too oversold on CIM as a cure-all to quality and operational problems, suboptimization will result.

10.7 Exercises

1. What are the two key ingredients in the Japanese methods of increasing productivity?
2. How is the Japanese style of finding out information different than ours?
3. Which comes first, product design or market research?
4. Who decides on product specifications?
5. Comment on why American managers do prototype work concurrently with process design.
6. How do our managers' pay structures relate to attitude toward productivity?
7. What is suboptimization?
8. What is the key to JIT/CIM implementation?
9. How do the stockholders fit into JIT/CIM implementation?
10. Which management group in a company is most resistant to the implementation of robotics?
11. How does the risk of the project influence a middle manager's thinking?
12. Why is the middle manager between a rock and a hard place?
13. Why do Americans tend to overdesign?

14. Why are the Japanese beating us in our own marketplace?

15. What does quality mean?

16. What does KISS mean, as applied to conformance to specifications?

17. What have some of the experiences of General Motors and Japanese firms taught us?

18. How does Pareto's law fit into JIT/CIM implementation?

19. How do we ensure that things happen as they should in terms of gathering information?

11

Integrating the Technology

11.1 Introduction

So far we have discussed all the various components that go into a CIM operation. Some of the technological and management ideas are well known, but most are still being evolved. Throughout this book we have stressed the idea that CIM is in reality a group of technologies, some of which may be more applicable in a given circumstance than others. Also central to the theme is the idea that for each individual component to be a part of CIM it must be integrated. CAD is a very interesting concept in that it allows the designer to employ the computer to assist in the design of things. However, until CAD is connected to CAM, we cannot say that either CAD or CAM is integrated, and hence factories that use CAD and CAM without the integration or connection feature are not CIM factories. So the key ingredient is, again, to integrate the various features. In this chapter we will look at the integration of the hardware or technological aspects and discuss the things that are required to integrate the factory to make it a CIM factory.

11.2 Standards

► ***Standards Are the Basis of Communication.*** Before anything can happen, people have to agree on the standards that will be employed. Decisions need to be made first as to whose standards will be adhered to. For example, if we are going to make a series of tests on the tensile strength of a material, we need to agree on whose standards we are going to use throughout the organization. In the case of material testing, the standards used are those of the American Society of Testing and Materials (ASTM). But what if you are going to sell products to Japan or Germany? Every country has its own standards groups covering all kinds of standards specifications. The material standards testing in Japan and Germany is not controlled by ASTM, although their procedures are likely to yield the same results on a given specimen. However, it is not guaranteed that the results will be the same because their methods may differ slightly and so the results may be subject to question.

Key National Standards Groups

In the United States we have the National Bureau of Standards (NBS). The NBS is augmented by the American National Standards Institute (ANSI). They will come to agreement, more often than not, with similar organizations in other countries. Some notable standards that have been agreed to by the NBS and others are the metric units defining what a kilogram is and how long a meter is. When we get beyond these basic definitions on length, mass, and time, and the like, the derived standards and the method of testing become murky. This is because each country has its own standards group and all these organizations are political. If a certain standard formulated by the United States is accepted worldwide, this gives prestige, and jobs, to U.S. citizens. If the French, who are very active in the standards area, have their proposed standards accepted, this adds to their prestige.

At least three U.S. groups are concerned with electronic and electrical standards. There is the Institute of Electrical and Electronic Engineers (IEEE), the National Electrical Manufacturers Association (NEMA), and the Electronic Industries of America (EIA). Sometimes these people are in agreement about certain items and at other times they are not. For the most part, each group has carved out a particular technical niche or territory to which they lay claim. For example, if you were to buy a panel to put electrical things into, you might specify a NEMA 14 panel.

▶ ***Old Technologies Have Well-Established Standards.*** For technologies that have been established for a reasonable length of time, the standards are fairly well established, and there is not much argument about the results of a test if it conforms to some standard. In the case of tensile testing, the ASTM publishes a series of procedures in many volumes, and what you do is follow the established procedure given. A report might look like this: The tensile strength of this specimen was 36,000,000 psi (footnote ASTM, Chapter X, verse Y). The establishment of whose standards are applicable will help reduce ambi-

Types of Standards	*Organization/System*
Physical	IEEE, ASTM
Operating systems	UNIX, MS-DOS, PC-DOS
Languages	BASIC, FORTRAN, C
Software	Lotus 1, 2, 3, VisiCalc
Interfaces	IEEE, EIA
Networks	ISO, NBS, COS
Other	NBS, ASCII, ANSI, NEMA

► ***Figure 11.1*** Standardization

guity in dealing with members within the organization and without, such as vendors, customers, and the government.

► ***Standards Groups for CIM Are Beginning to Appear.*** For more recent technology, the situation is very murky regarding standards. As indicated earlier, industry leaders such as GM are starting to force the issue by designating whose standards they are going to use and urging others to join them. In the case of CIM and particularly with communications, it is essential that standards groups be developed. Toward this end, GM, Dow Chemical, and CitiCorp have joined the recently established nonprofit Corporation for Open Systems (COS) located in Alexandria, Virginia. The purpose of COS is to accelerate the incorporation of the Open Systems Interconnection (OSI) and Integrated Services Digital Network. It is the aim of this group to establish a consistent set of standards and test procedures. However, until such standards are in fact established, the inability to communicate is still the major problem. Some standards organizations are listed in Figure 11.1.

11.3 Operating Systems

Electrical and electronic standards are the first step toward integration. The mechanical and hardware items are in various stages of having standards established. In the case of MAP, we have the IEEE 802.4, which establishes the type of connector used in the system and certain other characteristics of the connection. This is fine, but it doesn't go far enough.

► ***Uniform Operating Systems Are Sought.*** Computer vendors must make a concerted effort to establish a small number of *uniform* operating systems. As indicated earlier, most IBM PC machines use MS-DOS. However, there are many different versions of this operating system and not all are compatible. Before we leap into selecting any operating system, we must first take a look at the method of operation within our plant and see if only one system will suffice. Then we need to see which systems are available now and in the future.

▶ ***UNIX Has a Slight Edge.*** One characteristic of any operating system is whether it will be used in a real-time mode or in a time-share mode. Certain systems do not lend themselves to real-time operation. MS-DOS is one of these, and even though it might be possible to incorporate an IBM PC-type system in the factory, it might not be a good choice if we have many highly distributed systems. Operating systems come and go, as do languages and just about everything else in the computer business. However, many knowledgeable engineers in the process industries, which have had longer experience with distributed systems, feel that the UNIX and XENIX systems are the favored ones. However, the important point is not which system or which very few are the best, but that a concerted effort be made through meetings with all departments to pick a program that will fulfill current and future needs.

▶ ***Machines Come with Whatever the Builder Wants.*** From our discussion it may appear that all the group has to do is to say, "We are going to adopt such-and-such," and it will be done. It is not that simple. As far as the computer and communications equipment is concerned, most factories are built in bits and pieces. Each machine tool, robot, CAD system, and the like, has a programmable device attached to it, which is supplied by a specific vendor. Naturally, the particular PD vendor felt his system was the best, and for the applications with which he was most familiar this was probably true. The problem comes when the user wants to use a different machine tool because it has certain operational and mechanical features that are not found on competing machines. Unfortunately, that machine will have a PD that is not like any others in the factory. Some years ago this was not too much of a problem because the machine-tool builder would put on any PD the customer wanted. Now, however, we find that the people who are building the machine tools are either owned by the PD manufacturing company or vendor alliances have been established. It is more difficult for the user to get a nonstandard PD with a machine without paying a very large premium.

Since virtually all machine tools, gauges, test stands, data loggers, robots, and other equipment in the factory are now equipped with some sort of PD, it is not realistic to assume any two will be alike. Older machines may have components that are not available on newer machines and vice versa. The microprocessor chip makers such as Intel and Motorola have done a good job of ensuring that every time they come out with a new chip design, programs designed for the older chips can be used on the newer ones. Also, there are universal programming languages such as MDSI's Compact II, but this requires a postprocessor.

▶ ***New PDs Should Be Compatible with Old Ones.*** The chief ideas here are transparency and upgradable compatibility. For example, when an attempt was made to upgrade a personal computer to make it into an enhanced version, it was found that a large number of unwanted characters were displayed when a word processor program was used. Before, the word processor had run well on the unenhanced version of the machine. One solution offered was to use a different operating system, but this would have meant redoing all the data files. The machine maker should never have offered a new version that would not work with prior programs, but they did. Nor is this situation unique. Nearly every PD maker has done this kind of thing when bringing out newer models. As a result, there are factories loaded with complex, stand-alone equipment with which it is difficult to interface other equipment.

▶ ***Cannot Integrate without Compatible Operating Systems.*** The choice of an operating system or systems is crucial to the integration of the factory. Time spent in working with all departments at all levels will be well spent. Once the decision is made, then the company is locked in. The chore of upgrading existing equipment and programs to run with the system chosen will be difficult. Black boxes are starting to appear on the market to make some of the older systems compatible. These devices are priced in the $1000 range, and so it may not be worthwhile to convert all the equipment. It might be cheaper to replace the machine. It is not likely to be very efficient anyway. Sometimes it will be possible to make software modifications. The extreme solution is to build a completely new plant from scratch. This is, in many instances, what the automotive companies are doing.

11.4 Languages

▶ ***There Are Too Many Languages.*** The academic community invents languages frequently. There are many reasons why, but usually the new language is needed because of new technology. There is usually no problem until that technology gets out into the factory. At least three languages are available for each of the machine tools, robots, programmable controllers, and more recently vision systems and other AI applications. In terms of dealing with computers, there are dozens of languages to choose from. One definition of perpetual motion would be to try and find out which language is best.

▶ ***Companies Should Pick a Few Standard Languages.*** We need to develop a standard language or languages within the organization for the same reasons that we need to develop standards on operating systems and everything else to do with CIM. Again, the central theme is to have a language within our environment that is understandable and transportable. All U.S. citizens are required to have a working knowledge of English. There are many things wrong with English, but we do manage to muddle through with it. The same argument can be made for computer languages. After careful study, pick the ones you feel most comfortable with and implement these as the official languages of the organization. It would be nice if the languages you pick are the most efficient but, even more important, they must be universal.

► ***FORTRAN, BASIC, and C Seem to Be Entrenched.*** Many of our universities stress FORTRAN in the engineering curricula. More recently we find that Pascal is being taught as it has certain tutorial advantages relating to structure. With the advent of the personal computer, we find BASIC is now being widely learned through formal courses or, in very many cases, it is self-taught. Out in industry, especially the process industry, which is farther along with computer integration than the batch-flow metal-working industries, we find FORTRAN and BASIC are more universally used than others. Many process companies have standardized on FORTRAN or BASIC, with BASIC being the most common. Both are considered by many experts to be cumbersome and/or slow. It is difficult to operate in real time with BASIC, and although FORTRAN is better in this regard, it is considered to be more machine specific and is thus not particularly transportable. They both have the virtue that there is a very large body of hardware and software support. To get around the problems with FORTRAN, several large process companies are switching to C and are adopting UNIX operating systems. Some companies are seriously looking at Pascal. Although Pascal is gaining acceptance in engineering schools for noncomputer science majors, C is virtually unknown in academia by anyone but the specialists. Since the engineering schools are not offering C, many people will have to learn this language on their own or through company-sponsored workshops.

It could be well argued that languages should be left to the programmers because there are many user software packages from which to choose. There can be no real economic argument for reinventing a software program when one is available commercially. However, unless your only function is to be a button pusher, there will come a time when you need a program that is faster to write yourself than it is to ferret out the right software package.

► ***Even Software Should Be Agreed On.*** In the case of packaged software, spread sheets, data bases, and word processors are the most frequently specified. Here, again, it is necessary to set standards and to specify which system will be used. The situation is truly chaotic, because there is no relationship between the cost of a software package and its worth. Some cheap packages outperform some very expensive ones. Furthermore, many of the more expensive ones cannot be copied (not easily anyway), and it becomes very expensive to standardize on these. However, because the expensive packages are heavily promoted and because the makers provide technical assistance, which sometimes cannot be obtained from the vendors of the less costly programs, the number of customers is very large. One consideration in hiring clerical help is whether a person has been trained on one of the standard software packages and, if so, which one.

► ***Format All the Data the Same Way.*** Software packages generate data that are then manipulated and displayed according to the program set up by the software. If the group that establishes the standard software cannot come to agreement, it is at least essential that the data format between competing brands be identical. This will allow for the establishment of a data-base system where the data need be entered only once. A very large and needless expense in any organization is the practice of having many departments duplicate each other's efforts by continually reentering the same data. Even worse is to have departments enter different data for the same item. An example would be if sales and accounting have different dollar values for the same backlog.

11.5 Interfaces

► ***Specify Even the Plugs and Sockets.*** Interfaces include not only the plugs and sockets used to interconnect the various PDs, but also the formats of the data being transmitted. The manufacturing standards committees need to turn their attention to the physical hookup and how they are to be effected. If an RS port standard is to be established, shall it be RS232C or some other? What kind of coaxial cable will be used and what type of connectors are allowed on the standard? Will the data bits have parity and, if so, will it be even or odd? Some machines can only accept certain types. If this is the case, then the data need to be altered through software or black box means to massage the data into the proper format. If two-wire control loop transmitters for process control are needed, will the widely used 20-milliampere configuration or something else be used? The questions need to be asked and the agreements need to be ironed out. The goal is to have plug-compatible equipment, but first we need to know what kind of plug will be used. It sounds like a trivial problem, but when a piece of equipment is ordered with the wrong interface, the equipment has to be modified. The modification can be done either by the vendor or the user with the vendor's permission; otherwise the warranty is voided. It takes time to make even the simplest changes and it costs money, which is always directly or indirectly borne by the user.

11.6 Hardware

► ***Bring Your Suppliers into the Standards Discussions.*** As part of the planning process to bring a CIM environment into the production areas, consideration should be given to establishing hardware standards. This is very difficult to do because the hardware is controlled by the vendor. However, it may be possible to do a great deal through persuasion. An extreme example of persuasion is saying, "You supply the equipment to these specifications or you don't supply equipment." If you are General Motors you can do this and get away with it, as they did with the MAP protocol situation. However, the rest of the industry has to rely on less dictatorial means to achieve the desired results. One thing that can be done is to invite your leading suppliers to be a part of your standards teams. If it looks as though what they offer does not fit your standards, they will get the idea early enough to make the necessary alterations in their equipment. In the same vein, they can tell you if what you want is unrealistic. The standards groups of the various trade organizations in your industry can also provide help.

► ***You Can Always Default to IBM.*** After you have done all of this, certain decisions will have to be made regarding the specific hardware used, based on the standards established in your company. One decision that many companies have opted for is to select IBM and forget about the rest. IBM has all types of PDs and other equipment both for the office and factory. They also offer the SNA network to interface their computers. Unfortunately, not all IBM equipment is compatible with other IBM equipment, so there is still the need for interface boxes. IBM's technology is awesome, as is their customer support and marketing. However, some products are not as technically advanced as their competitors, and in some cases IBM has been known to wait until the particular segment of the industry is mature before they make changes. Very recently, IBM agreed to begin to offer OSI-based equipment in Europe. When the equipment becomes available, it will be very reliable and well engineered.

The single-source solution might be valid for a very small company or for a company starting out that has had no computers or automation. For the rest of the world, it is more realistic to try and make technological compromises as you go along. A company that has had a good history with Digital Equipment, for example, knows the personnel, and has worked with them in the past is not likely to dump Digital for a firm with whom they have had limited experience. Since all makers of equipment have a niche of expertise, it pays to buy the best equipment in the category and to resolve the differences in compatibility as the need arises. However, the situation regarding particular standards should be made very plain to all vendors as soon as the standards have been established.

11.7 Security

► ***Security Systems Should Be Uniform.*** As a part of the establishment of standards within an organization, there is the need to establish standards regarding security. As discussed earlier, we need security to ensure that data are not tampered with and information does not get into the wrong hands. The question of who has access to what is a policy decision made by top management. The usual foundation is on a need-to-know basis. If you do not have a particular requirement for information, then you should not be allowed to access it. Product design information is not usually relevant to someone in the payroll department, and pay information usually is not relevant to the designer. Who has access to what is frequently also organized along lines of position within the company. The CEO has greater access than the janitor.

► ***Data Encryption Can Cause Interdepartmental Problems.*** Sometimes a department encrypts their sensitive data. Many software programs will do this and many computers have either a built-in random-number generator or clock that can easily handle this task. The level of security here is not very good but is better than nothing. Again, different machines use different algorithms and have different generators, so what works on one machine may not be decipherable on another. This is true even if all the plugs, parities, and everything else is compatible. It would be frustrating and expensive to receive encrypted data that comes out of your machine as gobbledygook. The transmitting machine could decrypt the data before it is transmitted, but this would defeat a major portion of the purpose of encryption in the first place. A vulnerable spot in the system is the communication pathway itself. A simple wire tap is all that is required to get onto the system.

The group that establishes standards for security must make decisions as to what kind of equipment and routines will be used so that all machines can be fitted with the proper equipment. In a complex system, it is not unusual to have many different types of encrypters and decrypters as needs constantly change, people move about, and data are continually being classified and declassified. This situation is not desirable, however. As in any field, experts are available who can offer advice and solutions. Whatever is established, it should be done on a company-wide basis.

11.8 Summary

Corporations need to establish a set of standards to be used within their organization. Many industry associations, such as the National Electrical Manufacturers Association, are extensively involved in setting standards. Working both internally and externally to set standards will help reduce the interfacing problems with hardware, operating systems, data formatting and codes, languages, communications systems, and networks.

11.9 Exercises

1. Why are standards important?
2. What do we mean by setting standards?
3. Are standards groups political? If so, what is the consequence of that?
4. Name three different standards groups having to do with electronic products in the United States.
5. What does the Corporation for Open Systems do?
6. Why do we need operating systems and what are some advantages and disadvantages of existing systems?
7. Why are there many kinds of different hardware operating systems on PDs found in the factory?
8. Who has done a good job of ensuring that there is compatibility between older and newer systems?
9. What is the main design aim in going from old to new equipment?
10. Why is it desirable to have one or at most two standard languages in the factory?
11. Why not come up with one language that is universal and have that the best language?
12. Why should you learn a language since this is the programmer's domain?

13. Why do we need to specify software and why do some of the inferior software packages cost more than some of the better performing ones?

14. Why does the format of the data need to be standardized?

15. What are some of the ways you can implement hardware standards in your factory?

16. Where can you go for help for implementing standardization?

17. Why do we need to establish standards on a company-wide basis?

12

Getting the Most Out of the Employees

12.1 Introduction

► ***Top Management Support Is Essential.*** Nothing will work if the system lacks top management support and commitment. Since the incorporation of JIT ideas and the implementation of these ideas with CIM is a totally new and different concept to most top-line managers, they will have fears, doubts, and concerns. JIT requires that decisions be made by groups, and a high degree of meaningful employee participation is required from the lowest level in the organization to the top. Therefore, managers must make definite psychological compromises regarding their role in the organization. In a nutshell, top managers must be willing to assume risk. The group will make decisions that may not be in agreement with what the top managers think, but the top managers must go along with them. If a top manager countermands a well thought out decision made by a group lower in the organization, it will indicate that top management says they are supportive but, in reality, it is "conditions as usual." All the power is still being centered in the executive suite.

When business is good and the company is making a profit, it is easy for top management to loosen the reins and permit the lower levels to act independently. However, at the first sign of financial trouble, insecure top managers instinctively pull back all or most of the authority previously given. For JIT to succeed, the CEO must demonstrate through deeds that he or she is a true believer in participatory management. If the trust is violated just one time, the message will be transmitted, and regaining trust later will be extremely difficult if not impossible.

The remainder of this chapter is predicated on the assumption that management has made the commitment and has earned the trust of the employees.

12.2 Over-the-Wall Mentality

► ***Information Is Power.*** If we examine any pre-JIT organization, we will find a common thread. Every department or operating group does everything it can to hide information from other groups. Information is power. The individual's goal is definitely not the same as the goal of the organization, and most people pay only lip service to the "let's all pull together" propaganda that comes down from the upper levels. Each individual knows that pay is dependent on how well he or she compares with individuals in other groups. There is only so much money in the merit pay pool, which means that, if someone is to get above average pay, then someone else will get below average pay. This type of structure is not conducive to team play, the key ingredient in JIT/CIM.

► ***We Must Have Team Play.*** A lack of team play is clear in the traditional relationship between design engineers and process engineers. The design engineer gets the specifications from the customer and designs the product without any input from other members of the organization. The design is passed over the wall into the adjacent cubicle, which is occupied by the process engineer. His or her job is to figure out how to make the part as it is designed. This leads to inefficiencies if additional equipment and resources are required to make the part. Because the part may be difficult to fabricate, chances are the scrap and reject rate will be excessive. Bad parts also lead to customer dissatisfaction and field repairs. Furthermore, if the part is expensive to make, the price charged will have to be higher than otherwise might be the case. Customer dissatisfaction and noncompetitive prices lead to a loss of the market share, which, in turn, leads to recession and unemployment. Nor is the situation confined to process engineers and product designers. All groups and departments have similar communication difficulties and operate with the over-the-wall mentality. This kind of thinking should not be surprising, since we are accustomed to the factory system where work is broken down into its lowest components and one operation follows another.

► ***Over-the-Wall Mentality Is a Hindrance.*** Our first challenge is to destroy the over-the-wall way of doing things and to tie the various manufacturing functions together. The cartoon in Figure 12.1 conveys

► ***Figure 12.1*** Process and Product Engineering

this idea. In the case of the product–process design area, joint committees and group efforts must be started in order to integrate the functions so that process design is capable of making good parts and the product design group comes up with parts that can be fabricated. Except for specific considerations such as safety, the customer usually is not interested in how the part is designed or made. Even so, close cooperation with the marketing department is required in the product and process design stages to ensure that the product does what the customer expects.

► ***Job Rotation Broadens People.*** In addition to forming joint committees and holding meetings, some companies have their engineers rotate jobs so that the product designer of yesterday is the process engineer today, and vice versa. Extensive training in another's field helps a person to understand the other person's viewpoint. Some companies use the product manager approach: one person is assigned the responsibility for a product, including marketing, research, testing, fabrication, assembly, and customer acceptance. Often the product becomes a profit center. In this case, the success of the product will be the direct result of the success of the various participants. If merit pay is granted on the basis of group performance, then we have a powerful tool to ensure that the cohesive effort of the group outweighs the tendency of the individual to become a star at the expense of the group. In any case, the motivational force comes from recognition, particularly if the pay is adequate and fair. Being identified with a successful group enhances an individual's position within the company. This is what the Japanese do in their factories and it is a part of their culture. Some argue that they carry the concept too far and, as a result, the individual is buried by the system. Even so, we need to incorporate more of these kinds of motivational forces if we are to break down the over-the-wall barriers.

12.3 Individual Reactions to Integration

► ***We Need to Recognize the Individual.*** Contrary to popular belief, it is not the worker who is likely to offer the most resistance to the incorporation of JIT and CIM into the factory. The American worker, who many unenlightened people in our society have regarded as ignorant, stupid, and uncaring, is none of these. The problem is that often the worker is not treated with the respect due to him or her. The worker is not given proper information, motivated properly, recognized, humanized, or given the opportunity to express his or her individuality. The price we pay for this treatment of our single most important asset is loss of markets, loss of national pride, relegation to a debtor nation status, decline in our standard of living, and moving out of the top ranks of nations with highest per capita GNP.

► ***Make Quality Goods Inexpensively.*** The people who are no longer standing in unemployment lines because the benefits have expired and who are moving out of the "rust belt" know that flexible automation is coming and fundamental changes need to be made. Although not trained economists in any sense of the word, they know intuitively that their jobs were lost not through automation of their plant but by loss of market share because the company could not make good things inexpensively. If approached with dignity and careful planning, workers who have the good fortune to still have a job will embrace incorporation of change with open arms, particularly if they believe the changes will lead to job security and stability. They know that to have any job in the future they are going to have to make certain adjustments in their thinking. If the worker is approached properly, the transition to a CIM operation should be relatively painless as far as labor relations are concerned.

► ***Middle Managers Are Being Fired.*** The middle and lower manager, on the other hand, has rarely had the trauma of being out of work for extended periods. Until fairly recently, those laid off from one job usually found work within about 2 months. Most recently, however, white-collar workers and managers are being laid off in droves by major corporations. Office automation is blamed, but the real reason is that lower and middle management have outlived their usefulness.

This should not be surprising when we consider what middle management does. They collect information from lower echelons, interpret the information, and make expert recommendations to their superiors. They garner knowledge, distill it, and pass it on. From top management they receive instructions and pass on these instructions to the people who are doing the work. Once the instructions have been passed down, they check to see if the work was actually done according to the instructions. Consider then that these functions are exactly what a fully integrated computer system, complete with expert systems and other AI features, is intended to do. Some industry futurists are projecting that many of the skilled, semiskilled, and functionaries in the labor pool will be replaced with integrated systems. The typical manufacturing facility will then have a labor composition as shown in Figure 12.2.

► ***Organizations Will Have Fewer Layers.*** Corporations are finding that they can get along with five levels of management where before they had ten. Currently, 5 months is the norm for these people to find another job. They find themselves in the unique position of having to promote the CIM concept to top management, implementing CIM once they have sold it, and putting their careers at risk as a result. The blue-collar worker is given adequate training, is counseled, and is generally looked after in the transition because they always have the option to go on strike if unionized, or, if not, to become unionized and then strike. Not so with the white-collar people and lower echelon managers. No one speaks for them. As a result, very few companies attempt to train, counsel, or otherwise assist these people in the transition to a CIM operation and flexible automation. Lower-echelon managers have the most to lose with the implementation of CIM. Even if they are able to keep their jobs, it is likely that their power base will be greatly reduced.

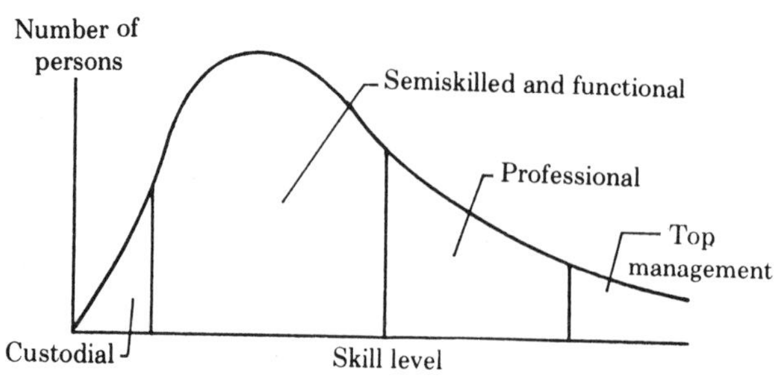

A. Before CIM

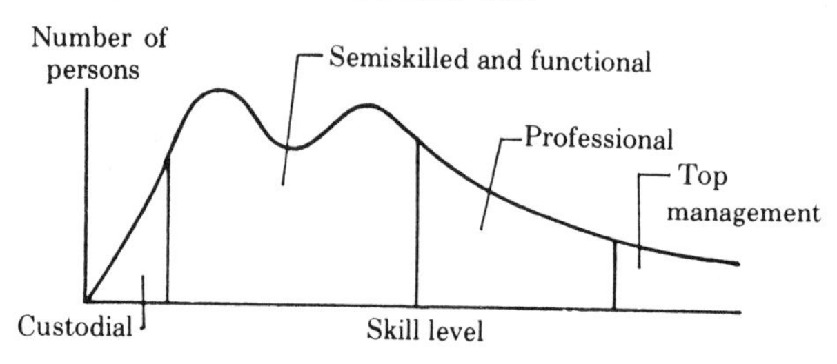

B. After CIM

► ***Figure 12.2*** Factory Employment

► ***Sell Top Management on CIM.*** What about top management? In nearly every recently published article on CIM, we see comments like, "We formed a study team and presented it to top management. Once they understood it, they really supported it." It appears that top management rarely takes the initiative, which makes you wonder what the basis of their salary is. After all, Deming and others have been telling us these things for 25 years, and the Japanese have been outdoing us in steel, autos, electronics, and appliances for 15 years. Most CEOs are now willing to listen because the concepts have been sanctioned in the *Harvard Business Review* and other prestigious journals. Once top management has been sold, they will learn about CIM and support the concepts. This is particularly true when business conditions have been generally favorable for a period of time and there is enough reserve buildup so that the companies can afford to take some risk and make the needed investment in the research and development. The first order of business if your operating department is in trouble is to sell top management on the CIM idea. Nothing is going to happen without their support.

► ***Different Levels Respond Differently.*** At the three levels, we find different individual responses to CIM. The CEO will actively and enthusiastically support it, or at least parts of it, when he or she understands what CIM can do. Middle and lower management and the white-collar worker will fight it if they feel in any way threatened. Those whose jobs are secure, such as the "knowledge" people (technical specialists who are irreplaceable), will go about looking for solutions to the problems. The blue-collar person on the factory floor will support the concepts if he or she is made a part of the decision and implementation process.

12.4 Skill Levels and Communications

Two things must happen if JIT/CIM is to be implemented. The skill levels of all employees and communications among people will both have to be improved. Skill levels is a function of training and education and more will be said about this later. Since training and education are inextricably linked with communication, it is communication that needs to be examined more closely.

▶ ***Improve Communications with Feedback.*** Communication is a process whereby information is *exchanged.* To communicate, the message must be transmitted, received, and understood. If you send a message to someone, that person needs to indicate that he or she received the message. A good way to ensure the message was actually received is to ask to have it played back. The next step is to ensure that the message was understood. If the person actually does what is asked, then chances are the message was understood as well. (Exchange is a result of the fact that you observe the actions and thus know the message was received and understood.) If feedback is not present, then the message may not have been understood and there has been no communication.

If you tell an employee he can't park in the executive parking lot and he does anyway, communication has failed. If you catch the employee and confront him, he may admit he knew he wasn't supposed to park in the lot. What happened is the employee failed to accept the message. Included in understanding a message is the idea that to be understood the message must be accepted.

Management may put an edict on the bulletin board, but if proper actions are not forthcoming the message did not get through and the manager has failed to communicate because the employee did not accept the message. This is called the acceptance school of management. It is essential, then, that there be provision in the interchange to ensure adequate feedback. Communication has the property of being similar to a closed-loop servo control system. If feedback is not present, there is no control. If feedback is distorted, the transmitter of the message must retransmit until the appropriate actions take place. The employee has to

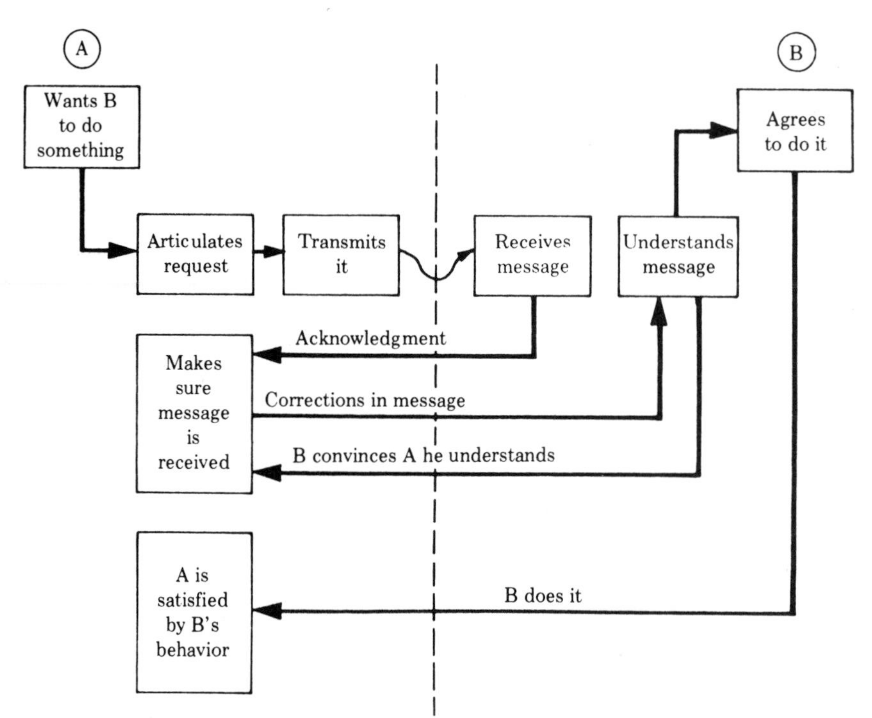

▶ ***Figure 12.3*** Successful Communication

understand and accept the message. If the employee does not do what is sought after repeated transmissions, the problem is likely a lack of clarity in the message being transmitted. This situation is shown in Figure 12.3.

▶ ***Messages Must Be Accepted.*** However, the employee may not be willing to accept the message for good reasons. In this case it is very likely that the employee is transmitting messages of his or her own and trying to tell the manager what is really taking place. If the manager is not listening and understanding, there is no communication. In the case of the parking situation, it may be that the employee lot is always full and the only available spaces are the managers' slots. The employee may find a reprimand less of a penalty than a reduction in pay for punching in late. For CIM to work, there must be an environment of true communication and communication must be both ways, top to bottom and bottom to top. The dictatorial manager who tells workers that "this is the word, take it or leave it" will likely find that they will leave it. The result will be that things will not get done the way the manager thinks they should be done and the project will probably fail.

▶ ***Mix with the Workers.*** What needs to happen if this situation is to be avoided is that the manager must get out onto the factory floor and listen to what the workers are saying. There needs to be a large psychological investment on the part of the manager and a willingness to spend time listening. Some doctors have learned that if they listen to a patient long enough the patient will tell them what is wrong. Everybody feels better, both patient and doctor have benefited, unnecessary tests and trial solutions have been avoided, and in the long run it is cheaper. Other doctors never listen to patients because they have the infallibility syndrome. They can do no wrong because they are doctors. These are the types who crash airplanes because they don't believe the instruments. The factory manager who has the infallibility syndrome will achieve results on the floor that are just as predictable. Getting out on the factory floor and listening to what the workers are saying is called "management by walking around," and it is one of the major points made in the book *In Search of Excellence.**

▶ ***The Medium Is Important.*** Many procedural things can be done to enhance communications. The medium itself is important. Sometimes it is advantageous to use the bulletin board. In many companies there is a computer bulletin board that is accessible to many of the employees. Meetings and training sessions can be used to exchange information, as can the in-house publication, called a house organ, that most large companies have. However, to be really effective, communications need to be on a one-on-one basis or in small groups. Frequent department meetings are an excellent means. Separate groups can be formed for other purposes and these bring together people with like interests. A good example are quality circles, which will be covered later. A special event, such as an awards banquet or a picnic, allows people to interact in a less formal manner and may open up channels to create more meaningful dialogues. Even the informal company grapevine is a powerful tool for management to communicate with the employees. The idea is to break down the "we" versus "them" barriers and to get people to think along the lines of "us."

12.5 Characteristics of the Job

▶ ***Good Job Descriptions Clear the Air.*** A careful analysis of what the people actually do now and what they will be doing in the future is required. Most smaller companies do not have written procedures for what is being done and most do not have job descriptions. Those that do will likely find that the written job descriptions do not reflect what the

*Peters, Thomas J., and Waterman, Robert H. Jr., *In Search of Excellence* (New York: Warner Books Inc., 1984).

worker is actually doing. The best job descriptions are the ones the workers write. It is a revelation to compare what the worker thinks he or she does with what the supervisor thinks, and both of these with what the personnel manual says the worker does.

▶ ***Put Content Back into the Job.*** In a CIM factory, the most apparent change will be in what the skilled worker has been doing compared to what he or she will be doing. When the machinist was in charge of the machine setup, doing fabrication and gauging of a part, the worker could see the results of his or her work immediately. This led to a great deal of pride in the work because the results were directly attributable to what the person did. As machine tools become more automated, individual achievement becomes less apparent. In a highly automated setup, the skilled craftsman is now simply a machine tender. Usually, setup, tooling decisions, and gauging are all done automatically. It is not uncommon now to walk through an American factory and see skilled operators keeping one eye on the machine to make sure nothing goes wrong and the other eye in a mystery story (unless the boss happens to be around). About the only functions these people serve are to make sure the tapes are running OK and to alert the appropriate expert if things go wrong. The content of the job has been downgraded over the years. Remember, we are dealing with bright and caring people. The older workers will ride it out as long as they can, and the younger ones will opt for going into other careers besides being skilled machinists. The ones in the middle will be thinking of retraining to move into more interesting and fulfilling jobs.

▶ ***Company Must Supply the Training.*** What about the bright and caring young people who will be moving into industry from our schools? If they are not receiving adequate education and training at the secondary and postsecondary levels, such as technical schools, industry will have to provide on-the-job training in the areas where they are needed. At a higher level we will find the "knowledge" experts. These people will be responsible for the care and repair of the complex machines. In many cases they will come from the ranks of the semiskilled and skilled worker who has been upgraded. Eventually, they will be the products of our technical schools and colleges.

▶ ***Managers Work Through People.*** People who have exceptional people-handling skills will eventually become managers. A manager is someone who gets things done through people. Some authors may argue that a manager can manage things, such as an advertising manager who places ads or a computer manager who runs the IBM mainframe. They may command the same salaries as people managers due to their competence and expertise, they may be called managers, and they may even have the key to the executive washroom. But if you do not have a "someone" who you are managing, you are not a manager no matter what you do. You can work with other people and interact with them to achieve common goals, in which case you are a facilitator or expeditor, but you are not a manager. A manager directs the activities of other people and has people reporting to him or her; the manager can fire, hire, or raise salaries. The courts agree to this definition and have confirmed the concept in cases involving workers who the fast-food corporations tried to exempt from the hourly wage laws. Calling a person who has no hire–fire authority a "shift manager" at a fast-food chain does not make that person a manager and does not exempt him or her from the hourly wage law.

▶ ***Expediters Are Especially Vulnerable.*** Since many people in lower and middle management are not managers, but facilitators and expeditors, and since these people provide functions similar to those a computer system and integrated data base provide, they are especially vulnerable. This is particularly true since in most cases they are not organized. Those who have people reporting to them, are effective leaders and motivators, and show consistently good results will always be secure. The job content will evolve to those who can manage people, those who possess specialized knowledge and expertise, and service people who will do the jobs that are uneconomical for the robot or machine tool.

12.6 Superior/Subordinate Relationships

▶ ***Practice Participation.*** We had alluded earlier to the idea of participatory management. This means that the manager must be willing to listen to the subordinate and to allow meaningful decisions to be made at the lower levels. Authority must be relinquished by the manager. Unfortunately, many managers will say that they believe in participatory management, go through all the motions of installing this type of management, with such things as various types of circles, and then not give the lower levels any real control over the budget. Employees will spot this phony type of game playing immediately and will be reluctant to cooperate in any future ventures. Managers must be willing to practice what they preach.

▶ ***Get Rid of the We–Them Attitude.*** Although professing to be in favor of participative management, managers will find that there won't be any real participation if they maintain the we–them attitude prevalent in American industry. In Japan, there are people at higher levels and people at lower levels, just like in the United States. However, nearly everyone in Japanese industry is considered to be an associate

and is spoken of as such. There is very much less of the superior/subordinate or traditional relationship of the "you do this or else" variety. There is more of "this is our problem and how are we going to solve it." Since the culture, structure, and educational forces in Japan are more directed toward the group as being the prime productive unit, as opposed to the individual, it may be that managers of Japanese factories have an easier time making the kinds of group decisions and achieving the participation required for JIT/CIM to function and grow.

▶ ***Japanese Methods Work in the United States.*** Contrary to what many people believe, it is not impossible for the Japanese style of management to flourish in this country. In the fairly recent past there were no Japanese companies operating in the United States. CEOs of U.S. industry, who were in competitive trouble, cited many things as being responsible for the Japanese success. These were cultural differences, a different work ethic, educational system differences, paternalistic corporate structure, restrictive tariffs, more favorable investment tax climate, better balance of trade, a higher personal savings rate, national policy for export by the government, and so on. However, we now have had several years of experience with Japanese companies that have moved into the United States with manufacturing operations. They are using U.S. labor, U.S. raw materials, U.S. capital equipment, and in many cases, through joint ventures, U.S. capital. Thus, these factories are using the same sources of supply that are available to U.S. manufacturers. These Japanese designed and managed companies are not at a competitive advantage on the basis of location and culture. Yet they are able to produce goods for less cost just as though the plants were located in Japan. The Mazda factory cited earlier is a good example. There have been others; most notable is the U.S. factory bought by Matsushta in 1976 to make electrical appliances. In this case the original factory was making TV sets and the company was not making a profit. Labor relations were poor and U.S. management wanted out and planned to move the manufacturing overseas. The Japanese came in and offered to buy the facility. They agreed to buy it under the condition that no employees could be laid off or fired. The Japanese were able to turn the factory around in about two years and make it a highly profitable operation that is able to sell goods competitively, and no one was laid off. In fact, because the operation has done so well, new people have been added. The argument of U.S. management has been proved specious.

▶ ***Openness Leads to Creativity.*** One problem Japanese management is encountering here is our openness and encouragement of the individual. Openness leads our people to be more innovative and creative. An effort is now underway in Japan to try and foster more individualism. This is proving to be as difficult for them as forming

14

meaningful group participation is for us. Japanese managers who have spent time in the United States and who have worked in our less structured environment find it very hard to readjust to the Japanese way of doing things. They are particularly frustrated because it takes so long for new ideas to be implemented. Apparently, there is a middle ground for which both cultures should be striving. Also interesting is that the government of Japan is now encouraging individuals to spend more, particularly on consumer goods.

12.7 Incentive, Merit, and Motivation

Motivation Methods

If we are going to implement JIT/CIM into our factories, we are going to have to alter the way we motivate people. It is assumed that the reader has had some exposure to organizational behavior theory, either through course work in schools or through outside reading. It will be recognized that the idea of participative management fits into MacGregor's theory Y scheme of things. Also useful is the general idea of Hertzberger that pay is a hygiene factor and as such is only a motivator, in the wrong direction if it is not adequate. Pay can be a positive motivator in the sense levels of pay denote a particular rank or status. Perquisites or perks are a very powerful motivator for the same reason. The corner office, the key to the executive washroom, the company car, and the like, are all provided to show that a particular individual is more important than the next person. The problem with all these things is they are individual oriented.

▶ ***Break Units into Entrepreneurial Cells.*** If we are to achieve group participation, we must look at ways to foster group effort and move individual effort into a more subsidiary role. The team versus the star. On the other hand, we do not want to make everything so democratic as not to provide any recognition to individual contribution. If we do this, we lose our stars and keep only the team players, which puts us into the Japanese dilemma cited earlier of losing the creative edge. Many companies, recognizing this dilemma, make it possible for creative individuals who possess sufficient entrepreneurial drive and temperament to work on their own, but within the corporate organization. New products are born this way.

▶ ***Visible Rewards Help.*** On the factory floor, where we are interested in producing quality products for less, we need to focus more on how to motivate each individual within the group. Public recognition of individual contribution is very effective in this regard. If someone comes up with an idea, enlightened management will make it possible for that person to get a significant amount of extra pay, either a raise or at least a bonus. Furthermore, if a group through joint effort makes the contribution, each member of the group gets like recognition. A public presentation of a medal or plaque by the CEO that can be shown to friends, with the person or group's picture in the paper, is long remembered. The presentation, award, or public recognition needs to have significance in order to have real impact. A $25 check as a result of a $100,000 cost-saving idea in the suggestion box won't suffice.

▶ ***Accounting Practices Have Not Kept Pace.*** There is a real problem in setting up a program that will reward the group effort. It has to do with the way we do our accounting in the factories. Traditional methods are geared toward keeping track of raw materials, work in process, and finished goods inventory as cost items. Associated with these costs is the direct labor of making the parts. Costing standards are calculated by various means, and deviations from these costs are recorded. When there was a very high direct-labor content in the cost of the goods sold, it was easy to see individual or even departmental contribution. Now direct-labor costs are usually less than 10% of the cost of the goods sold, and indirect costs along with administrative costs, which are not a part of the traditional cost of goods sold, are a very large portion of the value of the merchandise as it sits on the shipping dock. We need to develop accounting methods that reflect the true contribution of the elements that go into the fabrication of each individual product. The accounting profession is beginning to recognize and address the need to fit the accounting system to the realities of the flexible automated factory.

▶ ***Revamp the Suggestion System.*** Another condition that needs to be addressed is the U.S. idea of not rewarding people who are doing things that are a part of their normal job. In some factories the suggestion system pays for your suggestions for improvements in departments other than your own. The idea is that you are expected to make suggestions in your own department as a part of your regular effort. And it can be argued either way as to whether an intradepartmental award is beneficial. However, do you do anything for the machine repair person who comes up with an innovative way of getting a machine back on line an hour sooner when the line costs $100,000 per hour to run? True, fixing the machine was part of the regular job, but if management does not do something for this person, he or she will be thinking about moving on to someplace that will.

▶ ***Broaden the Personnel Function.*** Today the personnel department is called the human resources department, but traditionally their function has been to find and train people. Through industry associations, these departments keep track of pay scales and benefits and develop plans to generally keep people from being unhappy with the hygiene factors. Rarely do these departments decide who gets merit raises and who gets fired. Their function in most companies, although vital, is nevertheless still a staff relationship as opposed to a line relationship. Department managers still determine who gets paid and who stays. If we abolish the line relationships within the factory by incorporating groups and refer to everyone as an "associate," a managerial void will appear. Articles in the trade journals are beginning to suggest that the human resources department should fill the void and that it is the logical choice. To date, very few, if any, companies have placed this kind of line authority in these departments' hands.

12.8 Changing Group Dynamics

▶ ***Change Will Be Difficult.*** If we recognize the need for an alteration in the relationship between individuals within the JIT/CIM organization, we must also be aware that group dynamics will shift. These changes in group dynamics will not occur quickly, for there is at least 75 years of the superior–subordinate relationship that will have to change before the dynamics of the group will change. Management may say they are going to create an open environment and may show, by deed, that they are indeed committed to this objective. However, the people in the shop are naturally going to be suspicious of managment with a "born-again" attitude, particularly if the shop is highly unionized and if management hitherto has taken a hard line regarding the union.

▶ ***Share Financial Information.*** One thing that will help in achieving the transition is a scheme whereby the workers become part owners. Some companies have given their workers stock, some have given stock bonuses, and others have made it possible for the workers to take advantages of the same kind of stock-option privileges that are given to the people in upper management. Companies unable to give ownership have designed profit-sharing schemes. The money is important, but also very important is the sharing of the financial and marketing information (in simple, easy to understand language) so that the workers can see where the strengths and weaknesses of the operation really are and can then do something about it.

▶ ***Workers May Be Busy Making Scrap.*** Nothing upsets a line manager as much as seeing workers standing around idle. Perhaps the hardest thing these people are going to have to learn is that with the JIT mode of operation workers will have to stand around if the work flow is halted or if there are line imbalances. The idea is to keep the goods flowing at the highest useful productivity rate possible. This means making things for which there are firm orders and that can be shipped as soon as possible. If a manager does not accept the real goal and insists on keeping everyone busy, it is likely that the workers will all be very busy making scrap. Since the labor content in the part is low and the material content in the part is high, this approach is highly counterproductive. The workers are the first to know when the line is stopped and when the lines are imbalanced. If management encourages them, the workers themselves will figure out ways to keep the line in operation, if they are able to, and will work out ways to balance the lines.

12.9 Quality Circles

▶ ***Correct Implementation Is Essential.*** A great deal has been written about quality circles. The idea is simple enough: volunteer groups are formed to analyze the reasons why the quality level of that particular group has deficiencies and to design and implement corrective action. The group leader is elected by members of the group and, through the leader, information and quality concepts are communicated back and forth with management and other people directly involved in the quality programs. If, through the action of the group, the number of specification parts made increases while the number of out-of-specification parts decreases, a productivity gain has been made. The main function of the group leader is to ensure that the group is working on problems that they can possibly solve and that the meetings don't turn into gripe sessions. If a part from another department is defective, that information should be considered, but how to correct the defective part is the responsibility of the other department. It is not the responsibility of the user department. Note that this concept is directly opposite to the traditional U.S. suggestion system, which usually rewards people for making suggestions about operations in other departments.

▶ ***Group Cohesiveness Is a Result.*** When employee groups are formed whose actions have a direct bearing on the productivity, not only will productivity increase, but the groups themselves will become more cohesive. Farther down the line, the group members, either through their leader or as individuals, will go out of their way to cooperate with upper-echelon managers. Management can foster the building of group cohesiveness by allowing the groups to build group identity. Group names such as "The A Team" will pop up, bowling leagues will be formed, and members will meet after hours on their own time to discuss company problems.

▶ ***Management Earns Loyalty.*** The risk of group activity is that members of the group may become more loyal to the group than to the company. Very few workers in a traditional company have any real loyalty to the company. Most tend to feel that their loyalty account is evened up with each paycheck. They know that in difficult times their employment depends on their skill, seniority, attendance record, and productive output more than anything else. Even so, management should be aware of the feelings of individual members of the group and to try and direct some of the good will toward the company. If management treats the worker right and with the dignity, trust, and integrity he or she deserves, the worker will be loyal. Loyalty is something that management must earn, not something that the worker automatically gives.

12.10 Cost of Quality

Forming and nurturing quality circles will probably not solve quality problems nor lead to better products until a clear idea of what quality means and how to reduce its costs are fully understood. Clearly, the person most responsible for bringing the idea of cost of quality to the forefront is Philip B. Crosby, who wrote the book *Quality Is Free*.* Many companies have embraced his ideas, and there are now special seminars expounding the concepts and practices.

▶ ***Quality Is Conformance to Requirements.*** A customer ultimately sets a requirement upon which specifications are established and either the product (or service) meets the specifications or it does not. If it doesn't, nonconformance costs are associated with the product. Rework is a good example of this type of cost. Even if the product does meet the requirements, conformance costs are associated with the product. Routine measurements and tests would be examples of a conforming cost. If we identify all the costs, both conforming and nonconforming, and measure them, we can then work on ways to eliminate the nonconforming costs and reduce the conforming costs. A tabulation of cost of quality categories for manufacturing is given in Figure 12.4.

▶ ***Strive for Zero Defects.*** If a part does not conform, then it has a defect. If we eliminate the defect, we have eliminated the cost of the nonconformance. However, and this is the key point, to be successful we not only need to eliminate the defect but it is essential that we also prevent the defect from ever happening again. We strive for zero defects by preventing the defect from occurring in the first place. This means that we need to design, fabricate, and test the product correctly the first time we make it and everytime we make it.

▶ ***Defects Are Obvious in the JIT/CIM Factory.*** Notice how cost of quality concepts fits hand in glove with the JIT/CIM requirements and the Japanese way of doing things. If we devote the right amount of time and effort to designing the product so that it meets the customer's requirements and can be fabricated in our factory, then we should be in

*Crosby, Philip B., *Quality Is Free* (New York: Mentor Books, 1979).

Cost of Conformance	*Cost of Nonconformance*
Training: supervisor hourly	Rework
Special review	Scrap
Tool/equipment control	Repair and return expenses
Preventive maintenance	Obsolescence
Zero defects program	Equipment/facility damage
Identify incorrect specifications or drawings	Repair equipment/material
Housekeeping	Expense of controllable absence
Controlled overtime	Supervision of manufacturing failure element
Checking labor	Discipline costs
Trend charting	Lost-time accidents
Customer source inspection	Product liability
First piece inspection	
Stock audits	
Certification	

► ***Figure 12.4*** Manufacturing Cost of Quality (Compugraphic Corp., *Training Manual*)

a position to operate a smooth JIT operation. Any defect that does appear will be immediately corrected, as the production line will be stopped until the defect is found and fixed.

In a typical manufacturing operation, two-thirds of the total cost of quality might be nonconformance costs and one-third might be conformance costs. For a $100M operation, $66M could conceivably be saved by doing it right the first time. Incorporating a well-designed preventive program will certainly pay for itself very rapidly and will continue to bring rewards forever. Additionally, once people get used to identifying the quality costs in their department and eliminating them, they will continue to do so, particularly if there is an adequate built-in recognition and incentive system.

12.11 Retraining and Labor Relations

If a company moves into JIT/CIM, it must staff the factory with workers that have the necessary qualifications. To do this, the company has two major options. It can fire all the present workers and hire all new ones, or it can retrain the workers it already has.

► ***Replace or Retrain the Workers.*** The first method is not as far-fetched as it may first appear. We see this taking place all the time. Very few companies can get away with actually firing their present work staff and refilling the factory with new workers. The political and legislative uproar would be too great. More importantly, perhaps, the company may not be able to find workers in the area who are any more qualified than the ones they wish to eliminate. What happens frequently is that companies move to another location, say the Sun Belt or a developing country. A more subtle method is to phase out the manufacturing operation and become in essence a distributor. Many consumer electronics firms are in this position when they "private label" their TV sets and stereos.

From the viewpoint of manufacturing operations, there is little difference between actually firing everybody and moving the entire operation to another location and hiring new. This is still the preferred method of coping with the situation in many large corporations. The advantages are that the local union is eliminated, the capital equipment can be upgraded all at the same time, and willing and compliant workers can be hired. In labor-intensive industries, such as the shoe business, the main argument for moving was that wages were less in the South. This may have been true to a large extent in the past, but it is much less of a factor now. In the case of industries where JIT/CIM will work, we find, again, that the labor content is a minor part of the total cost. Other considerations, such as transportation, nearness to technology, and nearness to markets and raw materials, play a significant part in site location.

If there is a severe worker attitude problem in the existing plant due to prior management treatment of the workers, it will have to be changed before any retraining can be done. In many cases the costs of reshaping worker attitude and retraining may be the deciding factor on whether to stay or leave. However, if management treats its workers poorly in Detroit, what is to prevent them from treating their workers poorly in Greenville?

► ***Many Companies Opt for Retraining.*** The competitive advantage of moving south or overseas is less now than it was in the past. Also, stricter laws make it difficult for the plants to shut down without proper notice. In many cases, the union has followed the factories into the Sun Belt. For these reasons, many companies are finding it less costly in the long run to retrain their workers. Polaroid is an outstanding company in this regard.

► ***Productivity Brings Happiness.*** Contrary to what many people believe, happy workers are not necessarily productive workers. Hertzberg and other behavioral psychologists have shown us that unhappy workers are generally unproductive, but it is possible to have unproductive, happy workers. Later studies have shown that productive workers are generally happy workers, so we should strive to make our workers productive so that they might be happy. Some top managers will seriously question whether worker happiness is an issue. Remember, Elton Mayo gave us the coffee break because he demonstrated that productivity went up when workers took a break. Happiness was and is a nonissue in many managers' minds. Productivity is the key issue, and programs that improve individual productivity are mandatory. Fortunately, these programs also bring worker happiness in most cases, so they are socially redeeming and humane as well.

▶ ***The Union Wants the Same Things.*** The union is in favor of three things: better pay, secure jobs, and enhanced quality of the work life. Any program that promotes the well-being of the worker will be favored by the union. That is why unions are in favor of training (at the company's expense) and worker participation in deciding how the company will be run. Unions are not opposed to the use of robotics if the robots alleviate the drudgery and danger, but they are opposed to automation steps that reduce the number of workers in a factory. Although both union and management are in favor of enhancing the quality of work life and job content, they are so for entirely different reasons. Some of the major union contract clauses in a technological environment are shown in Figure 12.5.

▶ ***Missed Training Must Be Provided.*** We need to look at training from the viewpoint of what the company needs and what the employee brings to the company. In the case of robotics, automation equipment, and computers, the company needs factory people who can operate, install, program, and repair them. In general, this means the company is looking for people who are above average in intelligence and have had two years in a technological institute or college. If the individual is to hold a supervisory or managerial job, then perhaps the equivalent of four years is required. If the individual will be involved in design work, then engineering education is warranted.

For the most part, the person who is going to be doing repair work, for example, will need some technical training beyond the secondary level. Assuming the individual has this background, what will the company need to do to make that person productive? Factory automation equipment is made by many different machine builders. Each machine is, therefore, unique. For the operator or repair person to be effective, he or she will need training on the specific machine. This training must come from the employer. The training sessions can be in either the equipment maker's or the employer's facility, but ultimate responsibility for the training must come from the user of the equipment.

If the company is fortunate enough to have a technical institute nearby, then new graduates from that school can come into the company and will only need the machine specific or final training alluded to

Type of Provisions	*Specific Clauses*
Advanced notice provisions	Layoffs Plant shutdown or relocation Technical change
Interplant transfer and relocation allowance provisions	Interplant transfer provisions Preferential hiring Relocation allowance
Unemployment compensation provisions	Supplemental unemployment benefit plans Severance pay Wage–employment guarantee
Seniority and recall related provisions	Retention of older workers Merging seniority lists Retention of seniority in layoff
Exclusion from job security provisions	Exclusions from job security grievance procedure Exclusion from job security arbitration procedure
Work-sharing provisions applicable in slack work periods	Division of work Reduction of hours Regulation of overtime
Education and training provisions	Leaves of absence for education Apprenticeship On-the-job training Tuition aid for training
Provisions calling for joint labor–management committees	Industrial relations issues Productivity issues

▶ ***Figure 12.5*** Collective Bargaining Provisions and Technological Change (Ayers and Miller, *Robotics*)

earlier. However, technology is moving so fast that only very recent graduates will have enough background to be able to move directly into the final training programs. For older workers, the company can either provide in-house training of a more academic nature or they can send the worker back to the technical school for retraining.

▶ ***Basic Skills before Technical Training.*** In many cases, the individual may be able to cope with the additional training required and could pick up courses if he or she has sufficient background to handle the challenge. In other words, the person may be lacking in the basic math, science, and English skills, so nontechnical training may be required before the person can become eligible for the technical and final training required. This poses some questions for the company who is doing the hiring and retraining. For example, does a high school diploma qualify a person for moving into a potential technical job? Since the answer depends on what the person actually learned and not on whether the person got a degree, there is a need for extensive aptitude and intelligence testing as a part of the program. This is a serious problem in areas where there is a large minority population and is particularly severe in Spanish or other non-English speaking locations.

► ***The Customer Pays for the Training.*** We see that four levels of testing and training may be required before the individual can become productive in the CIM factory. This is shown in Figure 12.6, the triangle of learning. From an economic viewpoint, value has been added as the individual becomes more proficient in his or her function. Who is going to pay for the value added? One way or another, the company who hires the individual will pay for the value added. But, actually, it is the ultimate consumer who pays because the training costs will be included in the price of the goods sold. Some companies prefer to hire individuals who have been trained elsewhere, but these companies must be willing to pay a higher price for the individual. Other companies will conduct extensive training programs of their own and train the people they have who are capable of being trained or retrained. Many feel this is less disruptive as the person who is already working at the company knows the company technology, policies, and procedures. In many cases, if training at an outside school or university is the practice, the company will pay all or part of the cost of education. The individual will tend to have more of an obligation to stay when the training program is over. This is a sort of moral indenture. A common mistake many companies make is to train an individual and, when the training is complete, force the individual to leave by not paying competitive rates for the new level or by not giving the individual responsibility comparable to that of a new hire coming in at the same level.

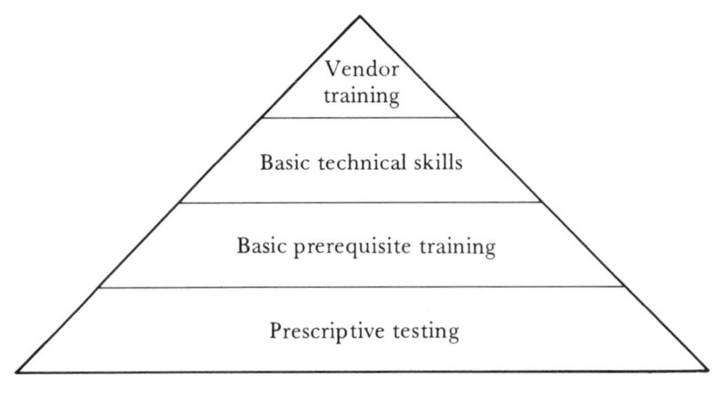

► ***Figure 12.6*** The Triangle of Learning (Malcolm, *Robotics*)

▶ ***Determining Qualifications.*** The current thinking is that if someone graduates from an engineering or technical school he or she is qualified. However, a great many incompetent engineers, particularly from countries whose educational system is not as good as ours, have found their way into our industries. It may be argued that, if a person can do the work and can demonstrate it by meeting the proficiency grade through a test, then that person should be recognized as being capable. Poor engineers, from whatever source, can be weeded out through proficiency testing, but for some reason proficiency testing is rarely used in this country in industry.

A notable exception is the programs promoted by the Society of Manufacturing Engineers. Certifications are awarded in many areas, such as welding, inspection, and machine operation. On the professional level, about the closest thing to proficiency testing is the testing for the registered Professional Engineer. Even though most companies in the private sector do not require the PE certificate or license, it is a good thing to have. More states are giving recognition to the PE status, it is required for many government projects, and it looks good on the résumé.

12.12 Displacement, Loss, and Grief

▶ ***Change Causes Stress.*** Some people can cope with change better than others. When we move a worker to another department or give the worker the option of going into a retraining program or being laid off, we create a very stressful situation for that individual. When we fire an employee, this creates extreme stress within the individual. Many companies are becoming aware of the stress process and are taking steps to lessen its effects.

▶ ***Terminated Employees Are on Their Own.*** In the case of the fired employee, usually nothing is done for him or her in terms of emotional and stress support, particularly at the lower levels. Termination benefits may range from nothing to a continuation of salary dependent on time of service. Frequently, outplacement services are provided at higher levels to help the employee find a suitable position elsewhere. These may include résumé writing assistance, introductions to specialist consultants, and the provision of services like a desk, secretarial assistance, and a telephone. Frequently, health insurance and other benefits are carried for a length of time after termination. However, once the decision has been made to get rid of the employee, the company's objective is to move him or her out as quickly and with as few repercussions as possible. Brutal and inhuman treatment frequently leads to lawsuits or at least makes it difficult to hire new people when the stories become known. Little if anything is done for the ex-employee's emotional well-being as this investment generally results in little payback or benefit to the company.

▶ ***Stress Leads to Low Productivity.*** For the employee who is to be retained but displaced to a new job, it is to the company's good to be concerned about the transition. A stressed and disturbed employee is not as productive as he or she could be, and here the goal is to bring that employee to a state of low stress as quickly as possible. Recent psychological and behavioral work has shown that there is a series of stages which a person must go through to get back to a condition of productive equilibrium. If the person gets hung up in one of the stages, equilibrium will not be achieved until whatever is causing the hang-up is resolved. If it is never resolved, the individual will never reach the desired equilibrium stage. At best, the employee will operate at a level where his or her inefficiency is not so bad you would want to fire him or her, but performance and productivity could be a lot better. The employee will always be marginal.

▶ ***Model Came from Hospital Studies.*** The original model for helping the individual cope was established by E. Kubler-Ross in her studies on death and dying. Many people who were terminally ill and many people who had suffered the loss of a loved one were interviewed. From this research came certain observations and conclusions regarding the emotional stages a person goes through when learning the ultimate bad news. Certainly, moving to another job within the company or even getting fired is nowhere near as extreme a situation as confronting your own death or the death of someone dear. However, there is enough evidence to conclude that the process of coping with the displacement situation is the same, even if the depth or degree of stress is much less. Both situations are concerned with loss and grief.

▶ ***The Stages of Grief.*** According to the model, there are five stages through which one must go to resolve the loss and grief. These are denial and isolation, anger, bargaining, depression, and acceptance. These steps are shown graphically in Figure 12.7. In the denial stage we say to ourselves, "This really can't be true," or "I'm not really going to be transferred to the shipping department." Our next stage is to become very angry. "Why those dirty so-and-sos." We can become angry at the situation, ourselves, or whoever is handy. Since the anger is a normal reaction, it needs to be brought out into the open. After anger comes the bargaining step. We may want to bargain with ourselves or whoever is in charge. The employee may do some extra work with the objective being "if my work is good then maybe the change won't be made." In the depression stage, the worker is likely to feel the world is against him, he is a failure, or the new situation will never be as good as the old. After the employee has gone through all these stages, he or she finally begins

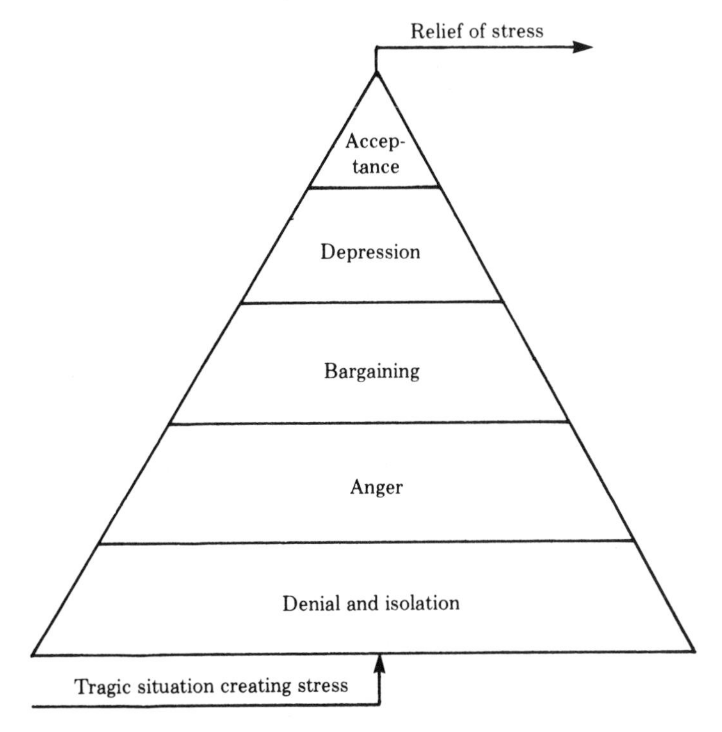

▶ ***Figure 12.7*** Grief Model

to accept the transition and will start to take on a more positive attitude. "Maybe it won't be so bad going back to school." The key point is that to get to the acceptance stage the other stages must be transited. Once acceptance is achieved, the employee will again be able to function normally.

▶ ***Techniques Are Not for the Untrained.*** The role of the industrial psychologist is to assist the worker in moving through the various stages. Often the individual will be able to work through the various stages on his or her own, but there may be cases where the person cannot do so. We have all met bitter people who are continually speaking against the company or who are quiet and moody. It is possible that these people are still hung up in one of the stages from a prior stressful situation. In any event, the transit can be made smoother and more quickly through the use of professional help. Whereas an individual may take six months to work through the process on his or her own, with help it may take half the time. The company must consider the cost–benefit ratio of getting professional assistance. The workers' associates can certainly help on a personal basis if they are aware of the stages or can get more qualified professionals to intervene in more extreme cases. Companies can help by providing stress counseling; relaxation exercises and physical exercise are frequently recommended.

12.13 Summary

Team play in a CIM facility is essential because without it the operation will not run smoothly, if at all. This is because in a JIT environment not only are people linked closer together in space but also in time. Unfortunately, political forces work to retard the team-play spirit. Most notable is over-the-wall thinking, which takes place between departments, say product engineering and process engineering.

Good communication is essential. As in any control-loop situation, feedback is necessary for communication to be effective. Not only must the recipient receive and understand the message, but he or she must also be willing to accept it.

People can be motivated by recognizing their individual and team contributions and forcing them to become involved in decisions and actions that affect themselves and their company. This adds content to the job that makes the job more meaningful. One way of involving people is to urge them to form and participate in quality groups. These groups are important because they let the members discover the problems affecting productivity and make it possible for them to come up with solutions and implement the solutions they have identified.

Two key areas that management will have to attend to are the training or retraining of employees and instituting programs to help employees alleviate stress due to the loss and grief that changes in jobs will cause. Extensive training inside and outside the company will be required to bring workers up to the skill levels needed to operate a CIM factory. Included are training programs so that individuals can operate advanced machines and also do maintenance and repair work on their machines, make improvements to their equipment, learn how to operate other machines, and perform quality assurance functions on the products being fabricated or assembled. Relieving stress is important because undue stress degrades productivity in an individual.

12.14 Exercises

1. What happens if a top manager loses the trust of his subordinates?
2. What is meant by over-the-wall mentality and how do we destroy it?
3. What are the advantages and disadvantages of job rotation?
4. What is one of the most important motivators as far as job performance is concerned for individuals and groups?

5. If the worker is approached properly, will the transition to CIM be smooth or not?

6. Why are lower and middle managers being fired from companies?

7. Who speaks for the white-collar worker and the lower-echelon manager?

8. How do different levels of management respond to CIM?

9. What is one way to improve communication?

10. What is the acceptance theory and what does it mean?

11. What is one of the key things a manager should do to find out what is going on?

12. What are some of the media that can be used in communication?

13. Why do we need good job descriptions and who should write the job description?

14. Who is responsible for supplying the training?

15. Why do the courts agree with the definition that a manager directs people?

16. Why must management be willing to practice what it preaches?

17. How do we get participative management and what prevailing attitudes do we have to destroy?

18. What is an associate in Japan?

19. What are some of the key people problems in a Japanese factory as far as creativity is concerned?

20. What are some of the problems with current pay structure and motivation?

21. What are some of the things we can do in terms of recognition and visible rewards?

22. In a modern manufacturing plant, what is the percent of direct labor cost to total cost?

23. What things can the human resources or personnel department do in a CIM environment?

24. Why is it important to share proprietary and other kinds of information with employees?

25. What does a quality circle do?

26. Who is the leader in a quality circle and where does he or she come from?

27. What is the main function of the group leader?

28. Why shouldn't the quality circle be concerned with implementing ideas in other departments?
29. What are some of the risks of excessive group activity?
30. Who is Philip B. Crosby and what has he done?
31. Why do we need to identify nonconformance costs?
32. What is the idea behind zero defects?
33. What happens when a defect occurs in a just-in-time factory?
34. What is meant by "do it right the first time"?
35. Why would a company want to implement extensive training programs instead of locating somewhere else?
36. Is a happy worker necessarily a productive worker? Is a productive worker usually a happy worker?
37. What is the union generally in favor of?
38. What kind of training levels will be required for different kinds of workers?
39. Who pays for the training of employees?
40. What is a Registered Professional Engineer and why should you be one?
41. Why should we be concerned about levels of stress in an employee?
42. What do we need to do to resolve the stress?
43. What are the five stages of grief?
44. Why do we need professional help to transit these stages?

13

Looking at the Future

13.1 Introduction

If I could foretell the future, I would own a seat on the stock exchange and enjoy my vacations on the Riviera. Even so, it might be worthwhile to speculate where CIM is going. If we look a year ahead, we are usually able to do better in forecasting than if we look 5 or 10 years ahead. If a soothsayer says the world will come to an end, he is correct. If the sayer tells you when, watch out! So, the predictions and speculations that follow are open ended. In that way, when questioned, I can always say, "Wait awhile and see."

▶ ***CIM Is an Infant Technology.*** Before we begin looking at the future, we should again review the perspective from the present. CIM and all related technology is in its infancy. There are fewer than 100 CIM plants in this country and less than 200 worldwide. Although many large corporations are supporting CIM, FMS, MAP, and all the other technologies in the alphabet soup, there is very little to look at by way of operational installations. There are several showcase operations, such as the Saturn plant, but these are few and far between, and those that are functioning properly are doing so with an elite crew of technologists and managers. These plants are not representative of typical factory conditions in that they employ better than average resources and most are intended as demonstration and R&D facilities.

▶ ***Pressures Will Force Change.*** Extremely powerful competitive forces will force changes both in hard technology (meaning machines) and soft technology (meaning management). Many companies are fighting for their survival as this is being written (e.g., Bethlehem Steel), and many other companies are steadily losing market share worldwide and will be in deep trouble soon if they do not take appropriate corrective action now. Fortunately, much of the technology needed is currently available, such as robotics and the JIT or Kanban systems, and other required technology is very nearly available commercially, such as AI and parallel processing. Economically, we have enjoyed about five years of prosperity (1987). There is enough of a capital reserve built up so that many companies can assume greater risks. With interest rates low, the cost of capital is attractive, which means hurdle rates are lower and paybacks are shorter.

13.2 Hardware/Software

► ***Advances in ICs Will Facilitate Implementation.*** Advances in integrated circuits (ICs) and memory storage will continue to bring down processing costs. Machines that currently have little in the way of automation will rely increasingly on ICs and will become more reliable and precise as a result. Custom or semicustom ICs will make it possible for machines that are unable to communicate with one another now to be able to do so. Integration of the factory will take place because the hardware will be cheap enough for a company to bring the machines together into networks.

► ***We Will See Many More AI Systems.*** Artificial Intelligence systems will grow very rapidly. We very badly need vision systems to guide our robots and we need good ways to reproduce the results of our thought processes. Heretofore the thrust on computers has been directed toward sequential processing. Our computers can do an excellent job in situations where there are few data but great output. Analysis and computational ability in humans is a left-brain skill and the sequential computer has moved into this area very well. The ability to make a judgment rapidly from a great deal of data, some of which may not be relevant (noise), is a right-brain skill. AI will be moving into this realm rapidly. One key to AI is the parallel processor. These are being built on a semicommercial basis using, essentially, discrete components. Again, as more IC parallel processors become available, the price of these computers will come down. All this activity will lead to machines that have more humanlike characteristics and, more importantly, they will be cost-justifiable, so they will get out onto the factory floor. Figure 13.1 shows how the technology choice is moving in this direction.

► ***Machinery Will Improve.*** Since the objective is to make machines that can operate at higher rates with precision over extended periods of time, the people who supply components such as motors, cylinders, and

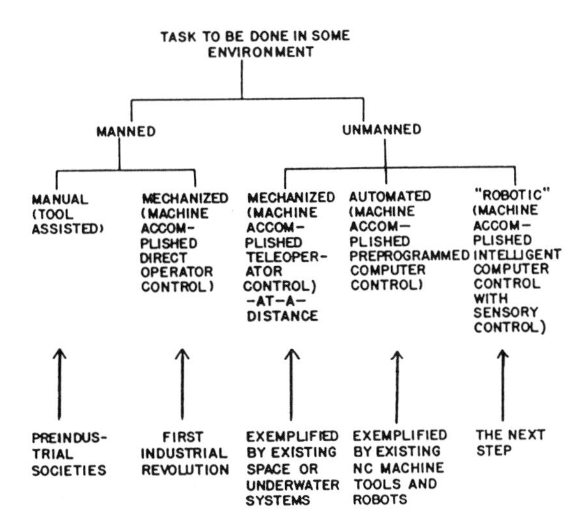

▶ ***Figure 13.1*** Technology Choice (Ayers and Miller, *Robotics*)

other mechanical parts will continue to be active in the area of making more reliable equipment. Certain technologies mainly used in the aerospace industry now will become more prevalent in the flexible automation arena. Metals can be plated with special coatings that greatly reduce friction, leading to machines that run faster with less down time. New materials such as ceramics and engineered polymers will be used in critical areas where traditional materials are insufficient. The kinds of activity required to bring these products and process along are just as high-tech as computers.

▶ ***Better Software Is Coming.*** Right now computer hardware is more advanced than the software. For this reason we see a great deal of activity, again in the AI area, to create software programs that will write software programs. This is essential if we are to be able to reprogram our robotic equipment quickly for the JIT operation where we are building products with batch sizes of 1 unit. There will be a continuing thrust to develop software packages that can be easily transported from one machine to the next, as well as programs that will be easily upgraded as newer machines become available.

13.3 Management Science

▶ ***More On-the-Job Training Will Be Provided.*** The real need in the management science area is to make the existing management science techniques available to everyone in the factory. This should be apparent if we recognize that in the CIM factory we will be dealing with groups of people all of whom will be participating in the decisions and implementations of the programs. If our "associate" is trained in group dynamics, organizational behavior, and related topics, communication will be facilitated.

We see today many training programs in statistical quality control methods and the conformance to specifications/zero defects philosophy espoused by Philip Crosby. In this latter context, if we are to have the worker become his or her own QA person and if we are to rely on the individual to do it right the first time, then these people need to be trained in these disciplines. Operations management programs, such as MRP, designed to assist the purchasing and control of inventory will become more prevalent. As these programs become more and more automated, people who once were doing data entry will be retrained into system maintenance and design personnel, in the same sense that the machine operator will be retrained to maintain the production equipment.

Current management science needs imply rapidly expanding education and training of all people at all levels. Not only will companies be compelled to provide paid time for the training, but additional resources such as training areas and training personnel will be required. The use of training aids, such as packaged video programs and specialized equipment, will be more common to facilitate the training and to make the process more efficient. As the market for training programs and equipment becomes larger, prices will decrease and quality will improve. There will be greater standardization of course content and methodology. The use of computers will increase, particularly when the computer training programs get away from the tutorial methods currently used and become more interactive. Again, AI techniques will be employed in this area and much research is being done in this regard.

13.4 Recap of Main Problem

There is really only one main problem. That is our tendency to parochialize everything we do. Parochialism means being in a state of selfish pettiness or narrowness. The problem manifests itself in many different ways in our factories.

▶ ***Main Technical Problem Is Lack of Compatibility.*** First is the situation regarding the lack of compatibility of programmable devices. For CIM to be successful, it is absolutely essential that programmable devices be able to communicate with one another. If General Motors had not forced the issue, chances are some of the larger companies would never have agreed to begin to make their equipment compatible and start to serve on the all-important standards committees. Each computer company thinks their own system is the very best. However, it is their parochial attitude that has caused the slow acceptance of their products in the marketplace.

▶ ***Must Eliminate Parochialism.*** We see parochialism in companies in the attitudes between various departments. The over-the-wall syndrome is a direct result of parochialism. Interdepartmental bickering and company politics wherein each person tries to gain advantage over the other leads to inefficiencies. In most cases, we can identify the parochial attitude of the various members as being at the root of the problem.

An attitude of "we're number one" can be healthy and beneficial as long as it does not interfere with the mission. Taking this attitude beyond certain limits can be counterproductive. Many individuals, as well as many companies, are in an unhealthy situation because they are full of the "not invented here" syndrome. If we didn't think of it, it obviously isn't any good. A sure way to lose market share is to assume that your technology is better than your competitors' and you know more than they do. We see this type of thinking on the national level as well. For years the Japanese have been doing things differently, but it wasn't until we started to encounter economic difficulties that we realized maybe they had some good ideas.

13.5 Possible Solutions

▶ ***Must Increase Market and Market Share.*** There are two ways to make any company grow. The first is to increase the total market and the second is to increase your market share within the total market. Since we are competing on a global basis, we ought to first consider the total market. It seems the first priority is to manufacture goods that can be used by people in developing countries as well as in the already developed countries. If we can build their economies through trade and other means, all producers will benefit. To achieve this, we must cooperate with other manufacturers of similar products to make it possible for people in less fortunate countries to buy the goods. Sharing of information and methods so that manufacturing generally becomes more efficient will bring prices down and make products more affordable.

To increase our market share, we need to make our factories more competitive by incorporating CIM and JIT thinking. We need to structure our organizations so that they allow for the exchange of ideas between those who set policy and those who carry out policy. Thus we need to instill a higher level of democracy and humanism in the workplace to break down the barriers of parochialism.

The real challenge is to convince our leaders and top managers that through openness not only comes happiness but also profits. They may not care about happiness, but they will be able to grasp the idea of profitability. The sooner you can educate them, the better for everyone.

13.6 Summary

There is great hope for the future of CIM although there are many problems in understanding and implementing it. Although certainly not a cure-all, CIM coupled with sound management practices will continue to grow. A major problem inhibiting the growth of CIM is the lack of compatibility at all levels in the technology. Parochialism needs to be destroyed before CIM can be implemented and this can be achieved through openness.

13.7 Exercises

1. Is it a good time to get into CIM?
2. What factor will help to bring the cost of CIM down?
3. What will happen in regard to expert systems and AI?
4. What is one of the key ingredients for success of AI?
5. What types of high technology will there be besides what is found in computers?
6. What do we foresee in terms of management science?
7. What is the main problem in the implementation of CIM?
8. What one thing must we eliminate in our society to achieve success in implementing any new technology?
9. What is the "not-invented-here" syndrome?

▶ *Bibliography*

Author Unknown, "The Reindustrialization of America," *Business Week*, June 30, 1980, p. 9.

Author Unknown, "ICAM Program Prospectus, Manufacturing Technology," *US Air Force*, September 1979.

Author Unknown, "A Report on Robotics in Japan," *Robotics Today*, 3, 26, Fall 1981.

Author Unknown, "Manufacturers Press Automating to Survive, But Results Are Mixed," *Wall Street Journal*, April 11, p. 83.

Author Unknown, "Tricky Technology: American Car Makers Discover 'Factory of the Future' Means Headaches Just Now," *Wall Street Journal*, May 13, 1985.

Abernathy, W. J., and J. M. Utterback, "Patterns of Industrial Innovation," *Technology Review*, June/July 1974.

Adlard, E. J., *Computer-Integrated Manufacturing—Its Applications and Justification*, Society of Manufacturing Engineers, Technical Paper #MS-394, 1982.

Albert, M. (assoc. ed.), "Computer Graphics: Cultivating the CAD/CAM Future," *Modern Machine Shop*, September, 1980, pp. 98–104.

Amrine, H. T., J. A. Ritchey, and O. S. Hully, *Manufacturing Organization and Management*, 4th ed., Prentice-Hall, Englewood Cliffs, N.J., 1982.

Baily, R., *Human Error in Computer Systems*, Prentice-Hall, Englewood Cliffs, N.J., 1983.

Banerjee, B. R., "Transitional Technology Transfer," *Journal of Metals*, January 1984.

Bateson, R., *Introduction to Control System Technology*, 2nd ed., Charles E. Merrill, Columbus, Ohio, 1980.

Battelle Laboratories, Columbus, Ohio, *The Manufacturing Engineer: Past, Present, and Future*, Society of Manufacturing Engineers, SME Special Report, 1979.

Blaesi, L., and M. Mannes, "Robots; An Impact on Education," *T.H.E. Journal*, March 1984.

Blum, A. J., "Computer Graphics—User's Report," *APEC Journal*, Summer 1976, pp. 8–10.

Boothroyd, G., C. Poli, and L. E. Murch, *Automatic Assembly*, Marcel Dekker, New York, 1982.

———, R. M. Milligan, and P. K. Wright, "Fault Decision in Manufacturing Cells Based on Three Dimensional Visual Information," *SPIE 1982 Proceedings*, Washington, D.C., May 1982.

———, and others, "A Flexible Manufacturing Cell for Swaging," *Mechanical Engineering*, Communications of the ASME, October 1982.

Brent-Davis, B. G., "Cam: A Key to Improving Productivity," *Modern Machine Shop*, September 1980, p. 106.

Bylinsky, G., "America's Best Managed Factories," *Fortune*, May 28, 1984.

Card, S., "A Method for Calculating Performance Times for Users of Interactive Computing Systems," *IEEE*, CH 1424-1/79/0000-0653, 1979.

Cole, R. E., "Target Information for Competitive Performance," *Harvard Business Review*, May–June 1985, p. 100.

Conway, R. W., "Priority Dispatching and Job Lateness in a Job Shop," *Journal of Industrial Engineering*, 16, 1965.

Cory, S., "A 4-Stage Model of Development for Full Implementation of Computers for Instruction in a School System," *Computing Teacher*, November 1983.

Crosby, P. B., *Quality Is Free: The Art of Making Quality Certain*, McGraw-Hill, New York, 1979.

Cross, R. E., *Increased Manufacturing Productivity through Education,* SME Manufacturing Engineering Education Foundation, Society of Manufacturing Engineers, Dearborn, Mich., 1981.

Dallas, D. B., "Special Report to Membership of SME, Society of Manufacturing Engineers, Dearborn, Mich., 1979.

——— (ed. in chief), "The Advent of the Automatic Factory," reprint from *Manufacturing Engineering,* Computer and Automated Systems of SME, 1980.

Davis, G. B., "CAM: A Key to Improving Productivity," *Modern Machine Shop*, September 1980, pp. 107–119.

Davis, L. E., *Design of Jobs* (Louis E. Davis and James C. Taylor, eds.), Penguin Books, Ltd., Middlesex, England, 1972.

Davis, R., and J. King, "An Overview of Production Systems," *Machine Intelligence 8*, E. W. Elcock and D. Michie (eds.), Wiley, New York, 1977.

Descotte, Y., and J.-C. Latombe, "GARI: A Problem Solver That Plans How to Machine Mechanical Parts," *Proceedings of the Seventh International Joint Conference on Artificial Intelligence*, Vancouver, B.C., Canada, 1981.

Division of Academic Planning, Office of the Chancellor, *Industrial Arts/Industrial Technology*, California State College and University, Fresno, October 1969.

Dixon, B., "Black Box Blues," *Sciences*, England, ca. 1984–85.

Driscoll, F. W., *Microprocessor–Microcomputer Technology*, PWS-KENT Publishers, Boston, 1983.

Droza, T. J., "The Challenge of Manufacturing Management," *Manufacturing Engineering*, September, 1983.

Drucker, P. F., "Goodbye to the Old Personnel Department," *Wall Street Journal*, May 22, 1986.

Dworkin, P., "Jobs in the '80's: A New Industrial Revolution," *San Francisco Chronicle*, San Francisco, September 24, 1982, p. 4.

Emerson, C., and I. Ham, "An Automated Coding and Process Planning System Using a PDP-10," *Computers and Industrial Engineering,* 6, 2, 1982.

Engleberger, J. F., *Robotics in Practice*, American Management Association, New York, 1980.

Erdilek, A., and A. Rapoport, *Technology Transfer as a Key Variable Affecting International Trade Competitiveness*, Division of Policy Research and Analysis, National Science Foundation, Washington, D.C., February 1983.

Esposito, A., *Fluid Power with Applications*, Prentice-Hall, Englewood Cliffs, N.J., 1980.

Evans, M., "Europeans Moving to CAD/CAM," *Modern Machine Shop*, September 1982, pp. 77–80.

Fadem, J., *Workforce Demographics and Their Implications for Human Resource Planning*, Center for Quality of Working Life, Institute for Industrial Relations, University of California, Los Angeles, February 1982.

Finch, W. J., "Why Aren't US Manufacturers Using Robots?" *Robotics World*, December 1983, p. 14.

Foston, A. L., *A Baccalaureate Degree Program Model for Computer Integrated Manufacturing*, Industrial Research Consultative Committee, California State University, Fresno, Calif., 1985.

Fox, M. S., "The Intelligent Management System: An Overview," *Process and Tools for Decision Support*, edited by H. G. Sol, Elsevier North-Holland, New York, 1983.

———, and others, "ISIS: A Constraint-Directed Search Approach to Job-Shop Scheduling," *Proceedings of the IEEE Computer Society Conference on Trends and Applications*, National Bureau of Standards, Washington, D. C., Technical Report CMU-RI-TR-83-3, Robotics Institute, Carnegie-Mellon University, Pittsburgh, Penn., 1983.

Freedman, M. D., "The Automated Factory in the '90's," CAD/CAM Technology, Computer and Automated Systems of SME, 1983, pp. 11–12.

Freiling, M. J., *Understanding Data Base Management*, Alfred Publishing Co., Sherman Oaks, Calif., 1982.

Fussell, P., P. K. Wright, and D. A. Bourne, "A Design of a Controller as a Component of a Robotic Manufacturing System," *13th International Symposium on Industrial Robots*, ASME, Chicago, April 1983.

Gatt, M. E., "Installing CAD/CAM: A Report from the Real World," *Design News*, March 26, 1984, p. 11.

Gettleman, K. (ed.), "Step by Step to the Automated Factory," *Modern Machine Shop*, September 1982, pp. 53–62.

Gevarter, W. B., *Intelligent Machines*, Prentice-Hall, Englewood Cliffs, N.J., 1985.

Gloess, P. Y., *Understanding Artificial Intelligence*, Alfred Publishing Co., Sherman Oaks, Calif., 1981.

Goetsch, D. L., *Computer-Aided Drafting*, Prentice-Hall, Englewood Cliffs, N.J., 1985.

Grandjean, E., and E. Vigliani (eds.), *Ergonomic Aspects of Visual Display Terminals*, Taylor & Francis, New York, 1983.

Grandjean, W. H., and M. Pidermann, "VDT Workstation Design: Preferred Settings and Their Effects," *Human Factors*, 25, 2, 1983, pp. 161–175.

Greene, A. M., "CAD/CAM in Europe; Autos, Aerospace Lead in Automation" (reprint), *Iron Age*, May 25, 1981.

Groover, M. P., *Automation, Production Systems, and Computer Aided Manufacturing*, Prentice-Hall, Englewood Cliffs, N.J., 1980.

———, and others, *Industrial Robotics*, McGraw-Hill, New York, 1986.

Harish, S., *Understanding Data Communications*, Alfred Publishing Co., Sherman Oaks, Calif., 1983.

Hayes, R. H., "Why Japanese Factories Work," *Harvard Business Review*, July/August, 1981.

Hearn, D., and M. P. Baker, *Computer Graphics*, Prentice-Hall, Englewood Cliffs, N.J., 1986.

Hedberg, B., and S. Jonsson, "Designing Semi-confusing Information Systems for Organizations in Changing Environments," *Accounting, Organizations and Society*, 3, 1, pp. 47–62, Pergamon Press, Elmsford, N.Y., 1978.

Hedges, C. S., *Industrial Fluid Power*, 3rd ed., Womack Educational Publications, Dallas, Tex. 1984.

———, and R. C. Womack, *Fluid Power Control—Electrical and Fluidic*, Womack Educational Publications, Dallas, Tex., 1983.

Hegland, D. E. (assoc. ed.), "CAD/CAM Integration—Key to the Automatic Factory," *Production Engineering*, August 1981, pp. 31–35.

Hertzberg, F., "One More Time: How do You Motivate Employees?" *Harvard Business Review*, January–February 1968.

Hodgetts, R. M., *Management: Theory, Process and Practice*, (3rd ed.), Dryden Press, Chicago, 1982.

Hoekstra, R. L., *Robotics and Automated Systems*, South-Western Publishing Co., Cincinnati, Ohio, 1986.

Hollomon, J. H., "Management and the Labor of Love," *Management Review*, AMA, January 1983.

———, and A. E. Harger, "America's Technological Dilemma," *Technology Review*, July–August 1971.

Hudson, C. A., "Computers in Manufacturing," *Science*, 215, February 12, 1982.

Humphries, J. T., and L. P. Sheets, *Industrial Electronics*, Breton Publishers, North Scituate, Mass., 1983.

Hunt, H. A., and T. L. Hunt, *Human Resource Implications of Robotics*, W. E. Upjohn Institute for Employment Research, Kalamazoo, Mich., 1983.

Jadrnicek, R., "Computer-aided Design," *Byte*, January 1984, p. 172.

Japan Industrial Robot Association, *The Robotics Industry of Japan*, Tokyo, 1982.

Jones, K., "Psst. Hey, Boris, You Want to Buy a Teensy Computer?" *Electronic Business*, May 1984.

Kaylor, D. (ed.), "WAM '83: Microelectronic Revolution Accelerates Engineering Shakedown," *Mechanical Engineering*, January 1984.

Keller, J. H., *The Management and Planning of Integrated CAD/CAM Interactive Graphics Systems*, Society of Manufacturing Engineers, Technical Paper #MS82-391, 1982.

Kinnucan, P., "Computer-Aided Manufacturing Aims for Integration," *High Technology*, May–June 1980, pp. 49–56.

Klein, B., "Productivity; A Dynamic Explanation," *Technological Innovation for a Dynamic Economy*, Christopher T. Hill and James M. Utterback, eds., Pergamon, Elmsford, N.Y., 1979.

Kosy, D., and V. S. Dhar, *Knowledge-Based Support Systems for Long Range Planning*, Technical Report, CMU-RI-TR-83-21, Robotics Institute, Carnegie-Mellon University, Pittsburgh, Penn., December 1983.

Kotkin, J., and Y. Kishimoto, "Theory F," *Inc.*, April 1986, p. 53.

Kroemer, K. H. E., "Ergonomics of VDU Workplaces," *Digital Design*, February 1983.

Krouse, J. K., "CAD/CAM – Bridging the Gap from Design to Production," *IEEE Transactions and Professional Communications*, PC-33, 191, 1980.

Kubler-Ross, E., *On Death and Dying*, Macmillan, New York, 1969.

Leep, H. R., and M. R. Wilhelm, "CAM on Campus: How Good Is It?" *CAD/CAM Technology*, Computer and Automated Systems of SME, Spring 1983, p. 12.

Lerner, E., "Computer-Aided Manufacturing," *IEEE Spectrum*, November 1981.

Lower, J. M., "Automation Heard Round the World," *Bobbin*, April 1985.

Lund, R. T., "Microprocessors and Productivity: Cashing in Our Chips," *Technology Review*, January 1981.

———, "Remanufacturing," *Technology Review*, February–March 1984.

Malcolm, D. R., Jr., *Robotics, An Introduction*, Breton Publishers, Boston, 1985.

Marsh, M., "Designing PC Boards in the '80s with CAD/CAM," *Design News*, February 6, 1984, p. 128.

Martelli, J. T., "The Need for a Contemporary Theory of Job Design," *Journal of Epsilon Pi Tau*, 8, 2, Fall 1982.

Maslow, A. H., "A Theory of Human Motivation," *Readings in Managerial Psychology*, Leavitt and Pondy, eds., University of Chicago Press, Chicago.

Merchant, M. E., "Delphi-Type Forecast of the Future of Production Engineering," *CIRP Annals*, 20, 3, 1971, pp. 213–225.

Mundel, M. E., *Improving Productivity and Effectiveness*, Prentice-Hall, Englewood Cliffs, N.J., 1983.

Nathanson, J. A., "How Much Education Is Needed to Become an Engineer?" *Technical Education News*, October–November 1981, p. 6.

National Science Foundation, "Manufacturing Employment Becomes Increasingly Technological," *Science Resource Studies Highlights*, February 23, 1983.

———, "Project Unemployment Scenarios Show Possible Shortage in Some Engineering and Computer Specialties," *Science Resource Studies Highlights*, February 23, 1983.

Office of Technology Assessment, *Exploratory Workshop on the Social Impacts of Robotics: Summary and Issues*, U.S. Government Printing Office, Washington, D.C., February 1982.

Ouchi, W. G., *Theory Z*, Avon Books, New York, 1981.

Petrosky, M., "Moving to Factory from Office," *Infoworld*, May 27, 1985, p. 34.

Porter, M. E., and V. E. Millar, "How Information Gives You Competitive Advantage," *Harvard Business Review*, July–August 1985, p. 149.

Potter, R. D., and R. Hinson, "Robotics Basics," *Robots 8 Conference Proceedings*, Detroit, Mich., June 3–7, 1984.

Puckett, A. E., "Productivity: A Management Commitment," *Hughes Aircraft Company*, 1980.

Quantz, P., *CAD/CAM—The Computer Invasion of Engineering and Manufacturing*, Society of Manufacturing Engineers, Education Reports ER81-02, 1981.

Ramalingam, P. R., "Computerized MRP System: Key to Profits and Productivity," *Modern Machine Shop*, September 1982, p. 66.

Reich, R. B., *The Next American Frontier*, Penguin Books, New York, 1984.

Reid, K. N., and others, "Trends in Mechanical Systems," *Mechanical Engineering*, March 1984.

Roberts, A. D., and R. C. Prentice, *Programming for Numerical Control Machines*, McGraw-Hill, New York, 1968.

Roman, D., and J. E. Pruett, *International Business and Technological Innovation*, Elsevier North-Holland, New York.

Royal Bank of Canada, "The Information Society," *Royal Bank Letter*, 66, 3, March/April, 1985.

———, "The Power of Recognition," *Royal Bank Letter*, September/October, 1985.

Savage, C. M., "Preparing for the Factory of the Future," *Modern Machine Shop*, October 1983.

Schmitt, R. W., "Technological Trends," *The Long-Term Impact of Technology on Employment and Unemployment*, National Academy of Engineering Symposium, June 30, 1983, National Academy Press, Washington, D.C., 1983.

Schoen, D. R., "Managing Technological Innovation," *Harvard Business Review*, May–June 1969.

Schonberger, R. J., *Japanese Manufacturing Techniques, Nine Hidden Lessons in Simplicity*, Free Press, New York, 1982.

Shackel, B., "Man–Computer Interfacing," *Human Factors Aspects of Computers and People*, Sijthoff & Nordhoff, Netherlands, 1981.

Shaiken, H., *Work Transformed: Automation and Labor in the Computer Age*, Holt, Rinehart and Winston, New York, February 1985.

Shapiro, B. P., "Can Marketing and Manufacturing Coexist?" *Harvard Business Review*, September–October 1977.

Simon, R. L., "CAD/CAM: The Foundation of Manufacturing Automation," *Commline*, May–June 1985, p. 14.

Singh, B., and T. L. Naps, *Introduction to Data Structures*, West Publishing Co., St. Paul, Minn., 1985.

Skinner, W., "Manufacturing—Missing Link in Corporate Strategy," *Harvard Business Review*, May–June 1969, p. 136.

Snyder, "Visual Ergonomics and VDT Standards," *Digital Design*, February 1983.

Snyder, W. E., *Industrial Robots: Computer Interfacing and Control*, Prentice-Hall, Englewood Cliffs, N.J., 1985.

Society of Manufacturing Engineers, *Increased Manufacturing Productivity Through Education*, SME Manufacturing Engineering Education Foundation, Dearborn, Mich., 1981.

Stammerjohn, E., Jr., M. Smith, and B. Cohen, "Evaluation of Work Station Design Factors in VDT Operations," *Human Factors*, 23, 4, 1981, pp. 401–412.

Tannenbaum, A. S., *Computer Networks*, Prentice-Hall, Englewood Cliffs, N.J., 1981.

Tanner, D. C., "Loss and Grief," *Journal of the American Speech–Language–Hearing Association*, 22, 11, ASHA, November 1980.

Tate, A., "Generating Project Networks," *Proceedings of the Fifth International Joint Conference on Artificial Intelligence*, Cambridge, Mass., 1977.

Taylor, A. P., *The Factory with a Future*, Society of Manufacturing Engineers, Technical Paper #MS82-399, 1982.

Topham, D. W., and H. V. Troung, *Unix and Xenix*, Brady Communications Company, Inc., Bowie, Md., 1985.

U.S. Air Force, *ICAM Program Prospectus: U.S. Air Force Integrated Computer Aided Manufacturing*, Materials Laboratory, Wright Patterson Air Force Base, Dayton, Ohio, 1979.

Upson, A. R., and J. H. Batchelor, *Synchro Engineering Handbook*, Muirhead Ltd., Kent, England, 1978.

Wilson, J. A. S., *Control Electronics*, Science Research Associates, Chicago, 1986.

Winter, R. E., "Computer-Guided Tools Are Catching On," *Wall Street Journal*, February 28, 1986.

Yen, D. W., and P. K. Wright, "Adaptive Control in Machining—A New Approach Based on the Physical Constraints of Tool Wear Mechanisms," *Journal of Engineering for Industry*, ASME, 105, 1983.

Acknowledgments (continued): **Figures 1.3, 3.3, 4.1, and 9.10:** Reprinted with permission from M.P. Groover et al., *Industrial Robotics: Technology, Programming, and Applications*. Copyright 1986 by McGraw–Hill Book Company.

Figure 2.1: Marvin E. Mundel, *Improving Productivity and Effectiveness*, © 1983, pp. 11–12. Reprinted by permission of Prentice–Hall, Inc., Englewood Cliffs, New Jersey.

Figures 3.1 and 9.18: Suri Harish, *Understanding Data Communications*. Reprinted by permission of Manusoft Corp., Culver City, California.

Figures 3.7, 4.11, and 7.1: Robert Bateson, *Introduction to Control System Technology*. Copyright 1980 by Charles E. Merrill Publishing Company. Reprinted with permission.

Figures 3.8, 3.10, 4.2, 4.3, 4.4, 4.5, 4.7, 4.8, 4.12, 4.17, 5.1, 5.4, 5.5, 5.6, 5.8, 5.9, 5.10, 5.12, 5.13, 5.14, 5.15, 5.16, 5.19, 5.21, 5.22, 5.24, 5.26, 5.27, 6.3, 6.10, 12.6: Douglas R. Malcolm, Jr., *Robotics: An Introduction*, Second Edition. Copyright 1988 by PWS–KENT Publishing Company.

Figure 4.6: Reproduced by permission. *Robotics and Automated Systems* by Robert L. Hoekstra. Delmar Publishers, Inc., Copyright 1986.

Figures 4.10, 8.13, 8.14, 8.15, and 8.16: Wesley E. Snyder, *Industrial Robots: Computer Interfacing and Control*, © 1985, pp. 18, 49, 50, 64. Reprinted by permission of Prentice–Hall, Inc., Englewood Cliffs, New Jersey. **Figure 8.13** also is copyrighted by Motorola Inc. Used by permission.

Figures 4.13, 4.14, 4.16, 4.18, and 4.19: James T. Humphries and Leslie P. Sheets, *Industrial Electronics*. Copyright 1986 by PWS–KENT Publishing Company. Reprinted by permission.

Figures 4.15, 4.20, 4.21, 6.9, and 8.2: *Robots and Manufacturing Automation*, C.R. Asfahl. Copyright © 1985 by John Wiley & Sons, Inc. Reprinted by permission of John Wiley & Sons, Inc.

Figures 5.3, 5.7, 5.25, and 9.14: Reprinted, by permission of the publisher, from *Robotics in Practice*, by Joseph F. Engelberger, pp. 24, 31, 44, 132, © Joseph F. Engelberger. Published by AMACOM, a division of American Management Association, New York. All rights reserved.

Figures 6.2, 6.4, 6.5, 6.6, and 6.7: Reprinted from G.C. Boothroyd et al., *Automatic Assembly* by courtesy of Marcel Dekker, Inc.

Figures 7.2 and 7.4: A.D. Roberts and R.C. Prentice, *Programming for NC Machines*. Copyright 1968 by Gregg Division, McGraw–Hill Book Company. Reproduced by permission.

Figures 7.7, 7.8, and 7.9: David L. Goetsch, *Computer-Aided Drafting*, © 1985, pp. 8, 34, 36. Reprinted by permission of Prentice–Hall, Inc., Englewood Cliffs, New Jersey.

Figures 9.4, 9.5, 9.6, 9.7, and 9.8: M.J. Freiling, *Understanding Data Base Management*. Reprinted by permission of Manusoft Corp., Culver City, California.

Figure 9.9: Paul Y. Gloess, *Understanding Artificial Intelligence*. Reprinted by permission of Manusoft Corp., Culver City, California.

Figure 9.11: V.W. Zue, *Tutorial on Natural Languages Interfaces, Part II: Speech*. Copyright 1982 by D. Reidel Publishing Company.

Figure 9.12: William B. Gevarter, *Intelligent Machines: An Introductory Perspective of Artificial Intelligence and Robotics*, © 1985, p. 48, 131. Reprinted by permission of Prentice–Hall, Inc., Englewood Cliffs, New Jersey.

Figure 9.15: Reprinted by permission from *Introduction to Data Structures* by Bhagat Singh and Thomas L. Naps, p. 112. Copyright © 1985 by West Publishing Company. All rights reserved.

Figures 9.16, 9.17, 9.20: Andrew S. Tanenbaum, *Computer Networks*, © 1981, pp. 9, 388, 396. Reprinted by permission of Prentice–Hall, Inc., Englewood Cliffs, New Jersey.

▶ Index